Smart Innovation, Systems and Technologies

Volume 91

The Smart Innovation, Systems and Technologies book series encompasses the topics of knowledge, intelligence, innovation and sustainability. The aim of the series is to make available a platform for the publication of books on all aspects of single and multi-disciplinary research on these themes in order to make the latest results available in a readily-accessible form. Volumes on interdisciplinary research combining two or more of these areas is particularly sought.

The series covers systems and paradigms that employ knowledge and intelligence in a broad sense. Its scope is systems having embedded knowledge and intelligence, which may be applied to the solution of world problems in industry, the environment and the community. It also focusses on the knowledge-transfer methodologies and innovation strategies employed to make this happen effectively. The combination of intelligent systems tools and a broad range of applications introduces a need for a synergy of disciplines from science, technology, business and the humanities. The series will include conference proceedings, edited collections, monographs, handbooks, reference books, and other relevant types of book in areas of science and technology where smart systems and technologies can offer innovative solutions.

High quality content is an essential feature for all book proposals accepted for the series. It is expected that editors of all accepted volumes will ensure that contributions are subjected to an appropriate level of reviewing process and adhere to KES quality principles.

More information about this series at http://www.springer.com/series/8767

Natalia Serdyukova · Vladimir Serdyukov

Algebraic Formalization of Smart Systems

Theory and Practice

Natalia Serdyukova
Plekhanov Russian University
of Economics
Moscow
Russia

Vladimir Serdyukov
Bauman Moscow State Technical
University
Moscow
Russia

and

Institute of Education Management
of the Russian Academy of Education
Moscow
Russia

ISSN 2190-3018 ISSN 2190-3026 (electronic)
Smart Innovation, Systems and Technologies
ISBN 978-3-030-08358-8 ISBN 978-3-319-77051-2 (eBook)
https://doi.org/10.1007/978-3-319-77051-2

Printed on acid-free paper

This Springer imprint is published by Springer Nature
The registered company is Springer International Publishing AG
The registered company address is: Gewerbestrasse 11, 6330 Cham, Switzerland

Preface

In 1937 L. Bertalanffy proposed the concept of a System and the development of a mathematical apparatus for describing systems. In 1970s A.I. Mal'tsev developed a theory of algebraic systems connecting algebra and logic for studying algebraic and logical objects. In 1990s the concept of purities by predicates was introduced by one of the authors and we found out some applications of this concept to practice. This conception based on the theory of algebraic systems allows to deep and clarify connections between quantitative and qualitative analysis of a system.

The book which is offering to you, "The Algebraic Theory of Smart Systems. Theory and practice", is an attempt to reveal the general laws of the theory of Smart systems with the help of a very powerful and expressive language of algebraic formalization and also an effort to use this language to substantiate practical results in the field of Smart systems, which previously had only an empirical justification. In fact, this is a translation of the theory of Smart systems from verbal language to a much more expressive language of algebraic formalization allowing in a different light to see the laws of the theory of Smart systems is proposed to the reader.

The key users of this book are persons which using elements of artificial intelligence in their work.

Moscow, Russia — Natalia Serdyukova
Moscow, Russia — Vladimir Serdyukov

Acknowledgements

We are grateful to Professor Lakhmi C. Jain for his proposal to write this book, for his constant help and support, vision and contribution.

Thanks are due to the publisher for assistance during the preparation of the manuscript.

Contents

About the Authors

Natalia Serdyukova is a Professor of the Academic Department of Finance and Prices of Plekhanov Russian University of Economics since 2012. During 1978–1997, she worked as an Assistant and then Associate Professor of the Academic Department of Higher Algebra of Moscow State Pedagogical University. During 1997–2012, she worked as a Head of Academic Department of Mathematics and during 2008–2012 as a Dean of the Financial Faculty of The Academy of Budget and Treasury of the Ministry of Finance of Russian Federation.

She received her Ph.D. degree in Mathematics at Moscow State Pedagogical University and Doctor of Sciences degree in Economics at The Academy of Budget and Treasury of the Ministry of Finance of Russian Federation. Now she has a title of Honored Worker of the Higher School of Russian Federation.

She has more than 70 publications in Pure and Applied Mathematics, in Finance, and in the Theory of E-learning.

Vladimir Serdyukov graduated from the Power Engineering Department of Baumann Moscow State Technical University and the Faculty of Applied Mathematics of the Moscow Institute of Electronic Engineering. He is a Professor at the Department of Applied Mathematics of Baumann Moscow State Technical University since 1997 and simultaneously Chief Researcher of the Institute of Education Management of the Russian Academy of Education since 2001. Doctor of Technical Sciences, Professor. His spheres of scientific interests comprise Operations Research, General Systems Theory, Probability Theory, Mathematical Statistics, Equations of Mathematical Physics, E-learning, IT technologies. He has more than 100 scientific works in these areas.

Introduction

The emergence of ideas and Smart technologies has changed the mentality of human society and, in particular, in the field of human communication, i.e., in the sphere of universal and public relations. This change is connected with the appearance of a more expressive language—the language of digits and digital technologies of building connections. At present, Smart technologies and Smart systems have become a common phenomenon in almost all spheres of human life. In 1937, Ludwig von Bertalanffy proposed the concept of a system approach and a General Theory of Systems and also the development of a mathematical apparatus for describing typologically dissimilar systems. His main idea is to recognize isomorphism, that is identity, sameness of laws governing the functioning of system objects. In the 1970s, A. I. Mal'tsev developed a theory of algebraic systems that connects algebra and logic and which is a universal mathematical apparatus for studying both algebraic and logical objects. In 1990s the concept of purities by predicates was introduced by one of the authors, and later on we found out some applications of the theory of purities by predicates to practice. This conception makes possible to get a new methodology for the study of systems theory based on the idea of formalizing a notion of a system using algebraic systems and methods of general algebra. It allows to clarify connections between quantitative and qualitative analysis of a system in order to specify the previously known concepts in the deepening of the study of qualitative properties. The book which is offered to you, "The Algebraic Theory of Smart Systems. Theory and practice," is an attempt to reveal the general laws of the theory of Smart systems with the help of a very powerful and expressive language of algebraic formalization and also an effort to use this language to substantiate practical results in the field of Smart systems, which previously had only an empirical justification. In fact, this book is a translation of the theory of Smart systems from verbal language to a much more expressive language of algebraic formalization. It allows in a different light to see the laws of the theory of Smart systems.

It is well known that the management uses the following abbreviation SMART, which has been proposed by G. T. Doran in 1981. The abbreviation SMART means a smart target and combines capital letters from English words indicating what the real goal should be: Specific, Measurable, Attainable, Relevant, Time-bounded. Thus, SMART is a well-known and effective technology for setting and formulating goals. The English word smart has the following meanings when translating it into Russian: intellectual, intelligent, reasonable, elegant, clever, strong, sharp, and some other meanings. If we ignore the generally accepted interpretation of the word "smart" as an intellectual, or the word "SMART", originally used in the management, then we can say proceeding from the generally accepted axiomatics of Systems Theory, that in fact any system in its functioning has one of its goals its highest, or, more precisely, the optimal level of development, that is, the level of smart. Recently, various scientific schools, for example, Russian statistics school, began to consider SMART as a fundamentally new social process. In this regard, the title of proposed monograph is as follows: Algebraic Formalization of Smart Systems, Subtitle: Theory and Practice. The subtitle is explained by the original purpose of the theory of systems, like any abstract theory—its ability to be really in demand and used by the human society. This key position for us largely determined the title of the book and its content.

Now let us give a brief content of the book.

We begin our considerations in Chap. 1 with three basic questions:

- what is the meaning of the concept of "formalization"?
- how to build a formalization, allowing one to obtain and justify meaningful result in the General Theory of Systems, not limited to empirical reasons?
- how to interpret the results of general systems theory to specific spheres of human activity, such as the general theory of training, IT technology, economy, and finance, as the link with almost all types of modern human activities (e-learning, IT, economics, and finance)?

The key point for us to answer these questions is the connection between logic and algebra. In 1986, I. R. Shafarevich observed that the present period of development of sciences is characterized by the mathematization of the science. Algebra has always occupied a leading position in mathematics. So, in Chap. 1 we shall consider the following main points:

1.1 The Concept of Formalization as a Tool to Study the Phenomena, Processes, and Practical Outcomes on a Theoretical Level
1.2 Two Directions of the Development of Logic. From Deductive Systems to A. I. Mal'tsev's Systems
1.3 Algebraic Formalization of the General Concept of a System, Based on Factors Determining a System
1.4 The Hierarchy of Algebraic Formalization
1.5 Probabilistic Algebraic Formalization

1.6 A Series of Distribution of a Complete Countable Distributive Lattice of Algebraic Systems. A Distribution Function of a Random Function of a Lattice of Algebraic Formalizations
1.7 Examples of Usage of Hierarchies of Algebraic Formalizations

The development of system approach and a lot of works devoted to the results in general system's theory brought up the question of what language these results should be expressed and how these results should be justified. System approach specifies General scientific methodology, so the justification of the results in this area should not be only empirical. More and more works of different complexity and different expressive means that offer various formal languages and approaches to describe the general system theory appear.

Here we give a commonly accepted intuitive definition of an abstract system used in many of the already known attempts to formalize a concept of a system. The system is the minimum set of elements connected by a certain structure which gives this set of elements certain qualities that ensure the achievement of the system's goal. In Sects. 1.1–1.3, we consider, from the position of algebraization of logic, the history of the question of the uprising of various formalizations in mathematics and show the transition from deductive systems to Malt'sev's algebraic systems in order to explain our idea of formalization. In Sect. 1.4, the expressive properties of various formalizations are considered and then a hierarchy of algebraic formalizations is constructed. In points 1.5–1.7 we defined the lattice of logics $\boldsymbol{L} = \langle L, \cup, \cap \rangle$ as a lattice of algebraic systems, concepts of a random formula, the probability of a random formula, the distribution function of a random formula, a random function of a lattice of algebraic formalizations in order to have an opportunity to monitor changes in a system's functioning.

In Chap. 2 we consider the performance of a system S by using an algebraic system A_S of factors determining the System S, P-properties of a system, where P is a predicate defined on a class of algebraic systems closed under taking factor systems and subsystems. The main points of this chapter run as follows:

2.1 Factors Determining the System
2.2 The Scheme of the Dynamic Predicates' Functioning in Models That are Groups
2.3 Cycles in the System's Development and Functioning
2.4 Algorithm for Determining and Regulating Smart System's Properties

We introduce in this chapter a method for modeling the final states of the system and determining the number of final states using the technique of group theory and the notion of purities by predicates or P-purities. Predicate P, i.e., function with a set of values from two elements $\{0, 1\}$ or $\{false, true\}$, that is, in fact, a condition that determines some property of the set separates the static properties of the system if it does not depend on time or on changing other external in relation to the system of factors. For example, for a class of algebraic systems, or for a class of all groups, or for a class of all abelian groups, the property of purity is a static one. Predicate P can single out dynamic properties of the system if it depends upon time or upon the changing of other factors external to the system. For example, if we consider financial systems and knowledge systems about them, then predicates that highlight financial sustainability, the legal sector of the economy, etc., are dynamic ones; that is, they depend on time, on the changing internal conditions of functioning of society. For learning systems, predicates that highlight levels of learning complexity which, in turn, depend upon the degree of development of society, and so on, are dynamic. Predicates, in contrast to numerical indicators, allow us to characterize the studied properties in a single integrated complex of both numerical indicators and links that are defined with the help of these predicates, and in the dynamic, if they are dynamic predicates, and in static if they are static predicates. An important question when studying the properties of a system and the process of its functioning, and in particular when studying the properties of a smart system and its functioning, is the question of how to determine that a system or a smart system ceases to satisfy some property P or some complex of properties Π. To answer this question, we introduce the notion of a partial probability measure on the set of unary predicates defined on the class of groups and closed with respect to taking subgroups and factor groups. In the point 2.2 in order to describe the change in the properties of the system during its operation and the possible change or adjustment of its target, we use dynamic unary predicates defined on the class of all groups. To characterize the functioning of dynamic predicates in models which are groups, we define a partial probability measure on the class of all unary predicates defined on the class of groups and closed with respect to taking subgroups and factor groups and consider binomial distribution for the realization of a complex of properties $\mathcal{P} = \{P_i | i \in I\}$ of a system S in n trials with the probability of successful realization of a complex of properties $\mathcal{P} = \{P_i | i \in I\}$ of a system S, equal to p, and partial binomial distribution for the realization of a complex of properties $\mathcal{P} = \{P_i | i \in I\}$ of a system S in n trials with the probability of successful realization of a complex of properties $\mathcal{P} = \{P_i | i \in I\}$ of a system S, equal to p. After that in points 2.3 and 2.4 we examine cycles in the system's development and functioning and construct an algorithm for determining and regulating smart system's properties.

In Chap. 3 we consider the simulation of the smart system with the help of finite group of factors determining the system, P-properties of the system, Cayley tables, and their role in modeling associative closed system with feedback. This chapter consists of the following points:

3.1 P-Properties of the Smart System. Sustainability of Smart Systems
3.2 Example. Smart Systems Modeling by a Group of Four Elements
3.3 Relationship between Factors Determining a System and Elements of a System
3.4 Substitution of Functions of a System. System's Compensational Possibilities
3.5 Compensational Functions of a Quotient-Flexible Smart System
3.6 Sustainability upon the Smart System Functioning
3.7 Loss Detection Point of Sustainability of a System Algorithm that Uses Models of Groups of Factors Describing the System

One of the most important issues of the Smart System Theory solutions for which the theory of finite groups can be used is a question about the sustainability of a system. Under the sustainability of the system we shall mean the system's ability to save its current state upon the influence of external and internal influences. Sustainability is a primary quality of any system. In the absence of this quality a system cannot exist. Sustainability brings together the various properties: resistance to external factors, sustainability, reliability, etc. In this chapter, it is from this position we shall begin to consider the question of the sustainability of the system which is defined in fact by the internal structure of its connections, robust, and interchangeability of structural resources. Any algebraic relations in the group $\boldsymbol{G_S}$ which is a group of factors determining system S defines communications in the system S and so the system's sustainability to some extent. If, for example, we have a system of equations

$$\mathop{\Lambda}_{i \in I} w_i(x_1, \ldots, x_{n_i}) = e,$$

then we can consider that it represents some connections between the elements of set of its solutions in $\boldsymbol{G_S}$.

Exactly from this position to study the property of sustainability of system, the notions of quotient-rigid and quotient-flexible systems are introduced. In the Chap. 3 with the same position we propose the following partial classification of the property of the sustainability of the system, which complemented the concept of P-quasi-sustainable system:

- the compensational sustainability or the factors' sustainability of the system S for the interchangeable factors a_i and a_j for the quotient-flexible systems,
- the sustainability with regard to the system's target of the system S which is described by the finite group of factors $\boldsymbol{G_S}$,
- the quasi-sustainability with regard to predicate which includes as a special case the sustainability with regard to the system's target,
- the final sustainability of the system,
- the compensational sustainability or the factors' sustainability of the system S for the interchangeable factors a_i and a_j for the quotient-flexible systems,
- the sustainability with regard to the system's target of the system S which is described by the finite group of factors $\boldsymbol{G_S}$,
- the quasi-sustainability with regard to predicate which includes as a special case the sustainability with regard to the system's target,
- the final sustainability of the system which we shall consider in the Chap. 10.

In Chap. 3 we also show that we can restrict the study of the infinite system by the usage of finite sets, namely finite sets of factors that determine the system.

In the Chap. 10 we shall establish a connection between the concepts of the final sustainability of the system and Lyapunov sustainability of the system.

After that in point 3.2 we consider an important question that arises during the study of the properties of the sustainability of the system is the issue about the possibility of mutual substitution of elements of the system, or the factors determining the system or the system's functions to achieve system goals. We consider this question in the present chapter. In order to outline possible solutions of this issue, we start firstly from the question about the relationship between the factors which determined the system and elements of the system, and then, on this basis, the issue of compensational properties of the system is considered. During this consideration, we outline the following main items: 3.2–3.7. In item 3.3 we show how one can use only finite sets to study infinite systems.

In the examples, we have considered the representation of a system S by a group of factors $\boldsymbol{G_S}$ where group $\boldsymbol{G_S}$ was finite. During the study system's properties directly the following question arises: How one can link the factors which determine the system and system's elements?

To solve this problem we shall use the following procedure.

Let system S consist of the following elements: $S = \{s_\propto | \propto \in \Lambda\}$.

Suppose that the group's of factor determining the system S main set is $G_S = \{e, a_1, \ldots, a_n\}$, and herewith the elements of a system S, defines each factor that is mutually relevant to each factor are set off:

$a_i \leftrightarrow S_i = \{s_{\propto_i} | \propto_i \in \Lambda_i, i = 1, \ldots, n\} \neq \emptyset$.

Note that each subset $S_i = \{s_{\propto_i} | \propto_i \in \Lambda_i, i = 1, \ldots, n\}$ of the set S one-to-one corresponds the subset $\{\Lambda_i | i = 1, \ldots, n\}$ of the set Λ.

The following cases are possible:

(1) $\{\Lambda_i | i = 1, \ldots, n\}$ is a splitting of a set Λ. This means that $\bigcup_{i=\overline{1n}} \Lambda_i = \Lambda$, and $\Lambda_i \cap \Lambda_j = \emptyset$ for any $i, j \in \{1, \ldots, n\}$ such that $i \neq j$.
(2) $\{\Lambda_i | i = 1, \ldots, n\}$ is not a splitting of a set Λ. By virtue of the definition of the group of factors which determine the system one can assume without loss of generality that $\bigcup_{i=\overline{1n}} \Lambda_i = \Lambda$. So there exist $i, j \in \{1, \ldots, n\}$ in this case such that $\Lambda_i \cap \Lambda_j \neq \emptyset$. In this case, we construct the grinding of the set $\{S_i | \mathrm{i} = 1, \ldots, \mathrm{n}\}$ up to the splitting $\{S'_\mathrm{i} | \mathrm{i} = 1, \ldots, \mathrm{n}\}$ of the set S, where $S'_i = S_i \backslash \left(\bigcup_{j=1, j \neq i}^{n} (S_i \backslash S_j)\right), i = 1, \ldots, n$.

If the condition (2) takes place then the intersection of clusters $S_i \cap S_j$ is called a reserve of functions f_i and f_j.

Conditions (1) and (2) lead to the following definitions.

Definition The system S is called a quotient-rigid one if the condition (1) is true. The system S is called a quotient-flexible one by factors and a_j if $\Lambda_i \cap \Lambda_j \neq \emptyset$, $i, j \in \{1, \ldots, n\}$.

The assumption of the finiteness of the group of factors G_S which determine the system S is not essential.

Thus, without loss of generality we can assume that each factor $a_i, i \in I$ from the group of factors G_S which determine a system S corresponds to a cluster of elements $S_i = \{s_{\propto_i} | \propto_i \in \Lambda_i\}$ of a system S. This correspondence is one-to-one.

In item 3.4 under the substitution (compensation) of a broken function of a system, we would understand the adaptation of the system to changing conditions of its existence and a replacement as a consequence of this broken or ineffective or not working elements of a system by relatively more efficient elements of a system. We would call such elements of a system as follows: substitutional elements or compensational elements. After that in item 3.5 we prove the following theorem.

Theorem *If the smart system S is a final sustainable one, then the elementary theory* $Th(\boldsymbol{G_S})$*, where* $\boldsymbol{G_S}$ *is a group of factors that determine the system S, is a complete one.*

Then in item 3.6 the question about how to describe the following situation in the functioning of a system is examined. Suppose that some non-empty set of elements of a smart system are out of order in the process of functioning of a system but, however, a system continues to function through its other resources and achieves its purpose. We shall offer the description of this situation for the case when the smart system S is described by a finite groups of factors $\boldsymbol{G_S}$. The next question which is considered in item 3.7 is the following. When modeling any system, the question of how to determine possible points of crisis in their functioning arises. The main question which arises here is the following one. Let $\boldsymbol{G_S}$ be a model of algebraic formalization of a system S not detecting or in other words not noticing the onset of the crisis point in the development of a system. The question of how we should change or supplement the model $\boldsymbol{G_S}$ in order that it would be able to predict or to "see" the onset of a possible crisis arises. The theory of catastrophes and the theory of bifurcations give an answer to this question for continuous models.

We propose the algorithm to determine possible points of crisis in our case, the case of discrete models of algebraic formalization of smart systems to use.

In Chap. 4 we consider the basic properties that determine the system: integrity, internal, and external attributive features that determine the system, that is allocating this system from all others. Then the integrity property is generalized to the case of P-integrity and P-internal and P-external attributive characteristics of the system, allowing to classify the properties of the system according to their various components. The formalization of the system goal made it possible to introduce the notion of a quasi-stable system with respect to the property P and the innovation system with respect to the property P. The essence of the system approach runs as follows: All the elements of the system and all operations in it should be considered only as one whole, only as an aggregate, only in interrelation with each other. Moreover, in constructing the algebraic formalization of smart systems, we shall fully take into account the Gödel incompleteness theorem, the essence of which is that it is impossible to describe the system by using the means of this system only. Therefore, to formalize smart systems, we apply a factor approach, corresponding definition in Chap. 4. Besides it, in this chapter we introduce the notion of external attributive features of the system and internal attributive features of the system, with the help of which we shall formalize the axiomatic of smart systems. As well a hierarchy of different levels links of the system is constructed. A theorem on the description of the system's links is proved. In addition, on the base of the theory of binary relations constructed by A. I. Mal'tsev, a classification of the binary relations of a system of each finite level is upbuilded. The obtained results are applied to models describing the system's synergistic effects and the processes of system's decomposition and synthesis. Chapter 4 consists of the following items:

4.1 Introduction
4.2 System Approach Basic Principles. System's Links. Connection with Synergetics
4.3 The Model of Hierarchy of Structural Links of the System
4.4 Types of System Connections. Different Types of Classifications. Classification of Binary Links of the First Level of the System
4.5 Closed Associative Systems with Feedback Partial Classification on the System Links Levels and the Number of Synergistic Effects
4.6 System Binary Links and Mappings
4.7 Algorithm of Analysis and Decomposition of the System by its Links Levels
4.8 Example. System Decomposition. Smart System THE World University Rankings. Evaluation of THE World University Rankings system
4.9 Algebraic Formalization of the Axiomatic Description of Smart Systems

The formalization of the axiomatic of smart systems we begin in item 4.2 with a review of the basic principles of the systems approach, the study of which requires the use of a synergistic approach. These are the following principles:

- the aggregate of the system's elements is considered as one whole, possessing a set of definite links and properties. So it turns out that the system is not a simple union of its elements. It is necessary to take into account the links between the elements of the system, providing certain properties of the system, that is the structure of the system;
- the properties of the system are not simply the sum or the union of the properties of its elements. The system can have special properties, which may not exist for the individual elements that arise due to the connections between the elements of the system, that is, due to the structural links of the system. The researcher, using a system approach, first decomposes the system into subsystems and elements, determines the goals of their functioning, the criteria for evaluating their effectiveness, builds models for their functioning, and then sequentially synthesizes them into the system model. This problem is extremely complicated, because the structure of system's links is not known a priori to the researcher. The structure of the system's links is closely related to synergistic effects. In item 4.3 we shall first concentrate on the hierarchy of the system's links. The main theorem of this item runs as follows.

Theorem about the description of the system's links.
Links of the level no more than n of the system S, where n is a natural number, are determined by no more than two combinations of connections of the level no more than n of the system S.

In item 4.4 we consider operations over system links and introduce a concept of a group of all links of the level n of a system S. After that in item 4.5 we noticed that the construction of an exhaustive detailed classification of closed associative systems with a feedback even over the levels of the system's links is hardly possible at the present time. Therefore, we consider here a special case. A partial classification will be made on specific examples which show how one can act in the general case within the framework of the assumptions made. Let us make the following remark. It is possible to classify the finite models $\boldsymbol{G_S}$ of factors which determine the system S, in the case when for each positive integer n the links of the system of level n has a finite group structure, that is, $\boldsymbol{C_n(S)} = \langle C_n(S), \circ, \square^{-1}, e\rangle$ is a group, since a complete description of finite groups has now been obtained.

Definition The system S is called factor-fractal by levels i,j, if the group of links $G_i(S)$ of level i is isomorphic to the group of links $G_j(S)$ of level j of this system.

Such a fractality we encounter, for example, in biology when transferring properties from parents to offspring. In item 4.6 it is shown that an important role in the process of decomposition is played by the split-off the links levels of the system, since the separation of links levels in the system is in fact decomposition. In 4.7 on the base of previous item, an algorithm of analysis and decomposition of the system by its links levels has been constructed. In item 4.8 we begin the investigation system decomposition on the example of the smart system THE World University Rankings. Smart system THE uses 13 parameters (or, evaluation criteria) with weights, which are expressed in percentages from the total score on the several categories of evaluation criteria. There are five categories (or, blocks) in this ranking system; as a result of previous theorem we get that this ranking is a rather sustainable one:

Let us consider a system $\boldsymbol{S}$ which represents THE World University Rankings. A decomposition of this system gives five subsystems, namely S_1, S_2, S_3, S_4, S_5; they correspond to each of the five mentioned above categories. Let G_S be a group of factors, which represent the system S. Let B_1, B_2, B_3, B_4, B_5 be respectively groups of factors, which represent subsystems S_1, S_2, S_3, S_4, S_5. We may apply the additional restriction on system S and subsystems S_1, S_2, S_3, S_4, S_5—the operation of composition of the factors is a commutative one. Under this restriction, a synthesis of system S is described by the following theorem.

Theorem *Let the operation of composition of factors which represent the closed associative system with a feedback be a commutative one. Then the synthesis of the systems* S_1, S_2 *is described by the group of factors* $Ext(B_2, B_1)$,[1] *the synthesis of the systems* S_1, S_2, S_3 *is described by the group of factors* $Ext(B_3, Ext(B_2, B_1))$, *the synthesis of the systems* S_1, S_2, S_3, S_4 *is described by the group of factors* $Ext(B_4, Ext(B_3, Ext(B_2, B_1)))$, *the synthesis of the systems* B_1, B_2, B_3, B_4, B_5 *is described by the group of factors* $Ext(B_5, Ext(B_4, Ext(B_3, Ext(B_2, B_1))))$.

[1] The group of extensions of an abelian group B_1 by the abelian group B_2.

Because the numbers of factors (that represent a close associative system S with a feedback with commutative operation of composition of factors) is finite, then, in this case—to some extent, there are some theorems which allow to simplify the synthesis process of the system S. In 4.9 the axiomatic description of the system and its formalization is given. Formalization of the system's properties using the predicates is considered also.

In Chap. 5 we consider the following questions: different approaches to the definition of duality in Smart Systems Theory, measurement of the system's links strength, the group of links of a system as a group defined on the Cayley graph of the system, the concept of efficiency and its formalization, the concept of *P*-efficiency of a system, *P*-subgroups of effective links of a system. Chapter 5 consists of following items:

5.1 Preliminary Facts
5.2 Several Examples
5.3 System Connections Strength. Example: The Social Relationships Strength
5.4 Duality in System Theory
5.5 The Connection Between Duality and the Concept of a Factor of a System
5.6 Algebraic Formalization of Modeling the Processes Preserving the Operation of Composition of Factors of a Closed System
5.7 Duality in the Theory of Strong and Weak System's Links
5.8 Efficiency (Utility of a Smart System). Formalization of Efficiency
5.9 Presentation of the General Task of the Smart System Effectiveness Determining in the Form of an Optimization Problem with Risks
5.10 Examples. The Use of Duality for Complex Smart Systems Classification by the Number of System Goals. Stability by the Parameter of Achieving the Goal of the System

In addition to Chap. 4 in Chap. 5 we shall continue to study the links of the system and define another group of system's links as a group defined on the Cayley graph of the group of factors that determine the system. Then we proceed to study the subgroups of effective connections of the system. We shall also introduce the notion of a common efficiency problem with risks. The next question that we shall consider in Chap. 5 is the use of duality in systems theory. The question of duality, and, in particular, the question of duality in mathematics, is one of the most interesting questions connected, notably, with philosophy. One of the methods for studying duality in systems theory is the tensor method of dual networks. We shall give a brief survey of both of these methods. In item 5.2 we outline here the following examples and their brief survey:

1. Connections in social systems.
2. Connections in physical systems, string theory.
3. Connections in the tensor method of dual networks.

In item 5.3 the links strength indicators introduced in that explain Granovetter's theory. In item 5.4 we consider other versions of constructing a duality theory for the theory of systems proposed. An important role in this matter is played by the formalization of the concept of the connection of the system and the clarification of its meaning. Here several ways of formalizing the links of the system are proposed. The first way to construct a formalization of the system's links runs as follows. The visual representation of the connections of the system uses graph theory. We have constructed on this basis, a group of the system's links that uses the Cayley graph of the group of factors G_S, determining the system S and the construction of the free product.

The second way runs as follows. Let the system link connects some elements a, b of the system, and we are examining the model of factors which determine the system S. Let this model be an algebraic system $\boldsymbol{A_S} = \langle A_S, \Omega \rangle$ of the signature Ω. The system's links should preserve but not destroy the internal structure of the system. So, it is natural to consider the homomorphisms of the system $\boldsymbol{A_S}$ into itself, that is the maps of the set A_S into itself, preserving operations and predicates from Ω, as the system's links.

Hence from we obtain several ways to study duality in smart systems theory.

The first way of constructing duality for the theory of smart systems uses models of factors that determine the system, and these models of factors are algebraic systems $\boldsymbol{A_S} = \langle A_S, \Omega \rangle$ of some signature Ω. Further the classical theory of duality from category theory is used in this method, [1]. It follows from the existence for each category the dual one that there works a duality principle in the category theory, that is, for every true sentence of the predicate calculus with respect to one category there exists a dual true statement for the dual category. The statement Pr^D, which is dual to the statement Pr and is formulated in the language of category theory, is obtained by interpreting in the category $\Re$ the sentence Pr, considered in the dual category $\Re^D$. A dual statement is obtained from the original one by preserving the logical structure of the statement and replacing in its formulation all the arrows by the opposite, and all products of morphisms into products of morphisms written in the reverse order.

The second method was proposed by us for the case when the model of factors is a group of factors G_S. Here we can consider the following two cases.

The first case: The group of factors which determined the system S is finite, and $|G_S| = n$. It is well known that in this case the group $\boldsymbol{G_S}$ can be embedded in the symmetric group of all permutations $\boldsymbol{S_n}$ of degree n. The second case does not use restrictions on the number of elements of the group $\boldsymbol{G_S}$. In the second method, we propose to embed $\boldsymbol{G_S}$ in its holomorph $Hol\boldsymbol{G_S}$. First of all let us consider the case where $\boldsymbol{G_S}$ is a finite abelian group. Then the holomorph $Hol\boldsymbol{G_S}$ of the group $\boldsymbol{G_S}$ is a semidirect extension of the group $\boldsymbol{G_S}$ with the help of its group of automorphisms $Aut(\boldsymbol{G_S})$. Let us use the multiplication form of the record for a group operation in $Aut(\boldsymbol{G_S})$, and for a group operation in $\boldsymbol{G_S}$ and in $\boldsymbol{Hol}\boldsymbol{G_S}$ let us use $\circ$ and +, respectively. The main set of the group $Hol\boldsymbol{G_S}$ can be considered as the set of all ordered pairs (g, φ), where $g \in \boldsymbol{G_S}$, $\varphi \in Aut(\boldsymbol{G_S})$. The group operation is given in $Hol\boldsymbol{G_S}$ by the rule: $(g, \varphi) + (h, \psi) = (g \circ \varphi h, \varphi\psi)$ for any $(g, \varphi) \in Hol\boldsymbol{G_S}$, $(h, \psi) \in Hol\boldsymbol{G_S}$.

In general holomorph of a group is the concept of group theory that aroses in connection with the solution of the following problem: Is it possible to include any given group $\boldsymbol{G}$ as a normal subgroup in some other group so that all automorphisms of $\boldsymbol{G}$ are consequences of inner automorphisms of this larger group? To solve this problem, we construct a new group $Hol(\boldsymbol{G})$, with respect to the group $\boldsymbol{G}$ and its automorphism group $Aut(\boldsymbol{G})$, whose elements are the pairs (g, φ), where $g \in \boldsymbol{G}$, $\varphi \in Aut(\boldsymbol{G})$, and in which the composition is defined according to the following formula:

$$(g_1, \varphi_1)(g_2, \varphi_2) = \left(g_1 \circ \varphi_1^{-1}(g_2), \varphi_1\varphi_2\right)$$

Herewith, the automorphisms $Aut(\boldsymbol{G_S})$ of the group of factors of the system $\boldsymbol{S}$ are in fact the links of the system $\boldsymbol{S}$ with special properties:

(1) the one-to-one correspondence between the factors that determine the system,
(2) the preservation by the link of the composition operation of the factors which determine the system.

In this connection, a special role here belongs to perfect groups, that is such groups $\boldsymbol{G}$ which are isomorphic to the group of its automorphisms $Aut(\boldsymbol{G})$. For example, $\boldsymbol{G} \cong \boldsymbol{S_n}$, where $n \neq 2, 6$. We have $Hol\boldsymbol{G}/\boldsymbol{G} \cong AutG \cong G$ for a perfect group $\boldsymbol{G}$. We obtain the following conclusion from all the above.
The main conclusion about duality.
The main idea of the representation of duality in the theory of systems runs as follows. Let $\{S_\propto | \propto \in \Lambda\}$ be a non-empty set of systems; and the element $a \in S_\propto, \propto \in \Lambda \underset{\square}{\Leftrightarrow} \propto \in \bigcap_{\propto \in \Lambda} S_\propto \neq \emptyset$. All such elements a give us in point of fact the set of all connections between systems $S_\propto, \propto \in \Lambda$. Elements of a new system S^d which is dual to the system S are the systems $S_\propto, \propto \in \Lambda$, that is the elements of the set $\{S_\propto | \propto \in \Lambda\}$.
In item 5.5 we return to Example 3 from Sect. 5.2, and namely to the definition of the tensor and the conjugate or dual vector space. According to the definition of the conjugate or dual vector space, we have $V' = \{f | f : V \to R\}$, where every f is a linear function from the vector space V into the field of real number R. Let us consider the correspondence

$$V' = \{f | f : V \to R\} \mapsto \{Imf \cong V/kerf \leq R\}$$

This correspondence, as well as the use of the concept of the group holomorph, helps us to introduce the concept of a factor system as a concept dual to the concept of a subsystem. This can be done for the model of algebraic formalization of the system by using the concept of factors which determine the system, and by using the concept of elements of the system for the system itself. In item 5.6 we consider several examples to clear up the process of decomposition of the education system by goals and links and after that the processes preserving the operation of composition of factors of a closed system and their formalizations. In the next item 5.6 we prove the following main theorem.

Theorem *Let S be a system and* $\boldsymbol{G_S} \cong \boldsymbol{V}$, *where* $\boldsymbol{V}$ *is an additive group of the Euclidean vector space of the dimension n, be a group of factors which determine the system S. Then the powers of links of the system S and the system S′ dual to S and defined by the group of factors* $\boldsymbol{V}'$, *where* $\boldsymbol{V}'$ *is an additive group of the vector space V′ which is conjugate or dual to the vector space V, are the same.*

In item 5.7 we introduce the definition of an efficiency function as an utility function. After that several examples of setting up the smart system efficiency function are considered. The next question that arises in the study of the utility function or the effectiveness of the system that we consider in item 5.8 is the question of how to link the effectiveness function and the effectiveness criterion with performance indicators. In item 5.9 we have noticed that the main conclusion about the representation of duality in the theory of systems allows us to tie up the proposed constructions to the classical approach of describing the properties of the system, which basically uses the notion of an element of the system, and not the factor which determines the system. This approach allows, for example, to classify complex smart systems *S* according to the number of goals of system *S* to study the stability of smart systems in terms of the parameter of achieving the goals of the functioning of the system.

In Chap. 6 we consider the following questions. We shall continue to study the concept of efficiency, and in line with this concept we introduce the definition of an innovative smart system, and in terms of algebraic formalization we shall describe the structure of innovative smart systems. To study the structure of the innovative system, the algebraic formalization of this concept will be used. In addition, we shall trace the analogy between the concept of an innovative smart system and the concept of an inverse limit, as well as analogues of the theorem on the description of abelian algebraically compact groups. Then we shall consider the concept of pseudo-innovative system, dual to the concept of an innovative smart system, and we shall get a description of pseudo-innovative systems using algebraic formalization. The notion of a quasi-sustainable system is introduced by the analogy with the concept of quasi-isomorphism from the abelian groups theory. An algebraic formalization of some properties of innovative smart systems and pseudo-innovative systems is constructed. Some examples of the use of these concepts consumed in the expert systems in training and in the economy are given.

After that we shall continue the empirical study of the process of system's decomposition using the example of the decomposition of the education system on the basis of these questions. The algorithm for a comprehensive assessment of the effectiveness of the functioning of the innovation system based on the tensor estimation of system performance is also proposed in Chap. 6. It is proposed to use homomorphisms of the group $\boldsymbol{G_S}$ of factors defining the system *S* into the group $GL(n, \Delta)$ of linear homogeneous transformations of the vector space R^n as tensor estimates of the efficiency of the functioning of the system *S*. One can also consider homomorphisms of the group of factors G_S that define a system$S$ in the group $GL(n, \Delta)$ of linear homogeneous transformations of the vector space Δ^n over an arbitrary field Δ.

The following results were obtained as applications:

- economic systems. In an economics with the presence of the shadow sector, a system with full implementation of P-connections cannot work autonomously if P does not implement any connections of the shadow sector.
- expert systems in learning. Testing with the full implementation of the links and the oral examination with the full implementation of the links give the same result. The levels of the impact of the knowledge system on the student have been singled out and tabulated. A more detailed consideration is given to one of the fragments of the knowledge system's decomposition, and it makes possible to determine the exposure levels listed in the table. Expert systems, an algorithm for compiling a database of errors, an algorithm for compiling a knowledge base, a theorem on describing errors, and a theorem on describing correct solutions have been used for this purpose. Together with the works, which deal with the issues of test quality and the practice of test assessment of knowledge in the Russian Federation, this makes it possible to determine the levels of knowledge of students with a sufficiently high degree of reliability.

Chapter 6 consists of the following items:

6.1 Formalization of Innovation and Effectiveness Concepts
6.2 Algorithm for a Comprehensive Assessment of the Effectiveness of a Smart System
6.3 Example. Decomposition of the Education System. Approaches to the Study of the Effectiveness of the Education System
6.4 Decomposition of the Knowledge System. The Representation of the System of Knowledge in the Form of an Algebraic System
6.5 Decomposition of the System. Analysis and Synthesis of the Knowledge Base

In item 6.1, we consider basic properties of innovative systems and examples of P-pseudo-innovative system. Item 6.2 is devoted to the task of constructing a numerical estimate of the effectiveness of the functioning of the system. This task is extremely difficult from a mathematical point of view since its solution involves a quantitative assessment of the appearance of qualitative changes.

We shall construct a tensor estimate of the effectiveness of the functioning of the system as a homomorphism of a group of factors G_S, determining the system S into a group $GL(n, R)$ of linear homogeneous transformations of the vector space R^n.

After that we construct an algorithm of a complex estimation of efficiency of functioning of the innovation system and the model of innovation management.

Then we consider the following questions: quasi-sustainability of pseudo-innovative systems, examples and application to expert systems in e-learning, economic systems.

In item 6.3, we decompose the education system into the following components: knowledge system S_1 (information subsystem), methodological and methodical complex S_2 of the knowledge system S (an adaptive subsystem), the system of students S_3 (target subsystem, target audience) and examine one of the fragments of the decomposition of the knowledge system in more details. In item 6.4 we consider the question of the representation of a system of knowledge. The brief survey of formal models of the representation of a system of knowledge and non-formal models of the representation of a system of knowledge (semantic, relational) is given. And item 6.5 is devoted to the question of decomposition of the system and to the question of analysis and synthesis of the knowledge base that arises in practice. In this item we construct the following algorithms: algorithm of errors description, algorithm for compiling a knowledge base, and search and analysis of correct task solving algorithms. After that we consider analysis of pupils solutions and transition to the record of the solution in the language of the narrow predicate calculus language (NPC) from the recording of the solution in the group theory language.

In Chap. 7 we have noticed that the accumulation of new properties of the system is associated with bifurcations or with the appearance of a qualitatively different behavior of the system element when a quantitative change in its parameters takes place, in accordance with works by I. Prigozhin and I. Stengers. It is assumed that the probability of reliable prediction of new properties of the system is small at the time of bifurcation, where bifurcation is a kind of a system regeneration. Contradictions arise naturally in the process of system's development, and they are the reason for the perfection development of systems. From the Theory of Systems, it is known that it is impossible to speed up the development of the system by artificially introducing contradictions into it, since it is impossible to determine whether the system, as a result of their resolution, will bear the new qualities. In this regard, one of the most important questions in the Systems Theory is the question of risks description. In this chapter, an approach to the classification of system risks from the position of algebraic formalization of the system is considered. This approach made it possible to distinguish between regulated (internal) and unregulated (external) risks of a system. As it is known, the question about the existence of infinite systems is debatable one. However, in this framework it is shown that the set of unregulated risks of any infinite system has a power of continuum accurate up to the regulated risks. An algorithm for managing the internal regulated risks of the system is constructed for a system represented by a finite group of factors. The risk function r of the system is defined as a function dual to the probability measure in the framework of algebraic systems formalization. This allowed us to consider probabilistic spaces with risk. Chapter 7 begins with an analysis of known existing approaches to risks description. In this connection, the main attention is paid to the quantitative definition of risk, which follows from the Kolmogorov–Chapman equation. Let us remind that the Kolmogorov–Chapman equation describes operations which occur according to the scheme of Markov random processes. Some relationships between the Kolmogorov risk function $h(x)$ and the risk function $r(x)$ introduced in Chap. 7 are found. Examples of the distribution functions $F(x)$ for which the risk function $h(x)$ is multiplicative one are considered. The risks of changes in formalizations ("failure of formalization") of the system using the

Kolmogorov–Chapman equation for exponential distribution are calculated. The statistical definition of risk is considered. Chapter 7 also presents a model of linear programming with risk. A linear programming model with a risk can be used in practice. Examples of the use of algebraic formalization for describing systemic risk in the particular case when the system of factors determining the risk of a closed associative system is considered in conclusion. In the case when the factors determining the risk of the system form a complete group of events that are independent in aggregate at any time and have the same probability density satisfying a certain condition, a measure of systemic risk is proposed. Chapter 7 consists of the following items:

7.1 Known Approaches to the Mathematical Determination of Risk. The Kolmogorov Risk Function
7.2 The Presentation of the General Model of Multi-criteria Optimization Problem in the Form of Linear Programming Task with Risks
7.3 System Approach to Risk
7.4 Mathematical Model of Risk
7.5 The Use of the Theory of Infinite Products to Quantify Risks
7.6 The Connection Between the Kolmogorov Risk Function $h(x)$ and the Risk Function r
7.7 Regulated Risks. Semigroup of Systemic Risks. Description of the System's Risk Semigroup
7.8 Algebraic Approach to the Description of Risks. Internal and External Systems Risks. Systemic Risk or System Risk

In item 7.1 we give a brief survey of well-known approaches to the quantitative definition of risk. In item 7.2 we present the general model of multi-criteria optimization problem in the form of a linear programming taking the risks into an account. In item 7.3 in order to ensure generality, we use an axiomatic approach to define the function of risk. In item 7.4 we construct risk function as a function dual to probability measure and examine its properties. In item 7.5 we consider some properties of infinite products that help one to quantify risks. In item 7.6 we consider the simplest examples of probability distributions with a multiplicative risk function. In item 7.7 we consider an external risk of the system, risks of formalization changes for the exponential distribution. In item 7.8 we introduce the notions of internal and external systems risks, consider the examples of risks functions on a finite σ-algebras, and construct an algorithm for regulating the internal risks of the system. After that we consider some properties of risk, prove the theorem about the description of systemic risk, and explore some examples.

In Chap. 8 we consider the question of how from an infinite model of factors that determined the system S one can go to the finite model of factors $\boldsymbol{G_S}$ which determine the system S. A list of necessary information from the finite groups theory, useful in the study of certain features of the functioning of the smart system, is given in addition. Table of classification of system's properties by models of finite groups of factors that determine the system in which some system's properties are classified is constructed on this basis by the models of finite groups of factors determining the system. The question about risk modeling in a smart university also is considered in this chapter. The model of an algebraic formalization of six factors of the risks of changes in long-term period of a development of the smart system is constructed on the example of the smart university. The algorithm of search of points of regulation of the closed associative system's functioning on the example of the model consisting of six factors is shown in this chapter too. Chapter 8 consists of the following items:

8.1 The Transition from an Infinite Model of Factors that Determine the System to a Finite Model of the System
8.2 The Necessary Information from the Finite Groups Theory Useful in the Study of Some Features of the System's Functioning
8.3 The Model of an Algebraic Formalization of Risks of Changing the Scenarios of the Long-Term Development of a Smart System of Six Factors on the Example of a Smart University
8.4 A Selection of Factors to Determine Long-Term Risks of a System
8.5 Conclusions. Future Steps

In item 8.1 the construction which is proposed helps one to create the finite model G_S of the system S in the form of a finite group of factors determined the system S. In fact to construct a Cayley table for the group G_S one can act in two following ways:

(1) To use combinatorial methods. One should search the defining relations of the model G_S with the help of simple enumeration.
(2) To make a qualitative analysis of the factors which determined the system and on this basis to explain the relationships between them.

In item 8.2 we give some known and interesting facts from the theory of finite groups that can be useful in studying such properties of the system as the presence of synergistic effects, the number of possible variants of forecasts for the development of the system, the stability properties of the system. The main idea of item 8.3 which unites the further presentation of Chap. 8 is to show that, with the correct and timely regulation the process of the system's functioning, it becomes a smart system in the sense of the optimal system on the selected smart criteria, or, in other words, a smart optimal system. In the proposed model of algebraic formalization of risks of changing the scenarios for the long-term development of a smart system of six factors, the risk of formalization's change from the symmetric scenario to the cyclic scenario, and the tensor index of the effectiveness of the system performance on specific indicators, can be calculated by the algorithm proposed in this section, for example, for a smart university system. In item 8.5 some future steps are proposed to develop a methodology of SmU modeling as a system based on an algebraic formalization of general systems' theory, theory of algebraic systems, theory of groups, and generalizations of purities, and identify formal mathematical conditions for a system—in this case SmU—to become efficient and/or innovative.

In Chap. 9 we return again to the special case in which the factors affecting the system determine the group. In this case, the system is a closed associative system with feedback. This chapter consists of the following items:

9.1 Particular Case: Factors Affecting a System Determine a Group
9.2 The Group of Automorphisms of the Group of Factors that Determine the System
9.3 Direct and Inverse Spectra of Groups and their Limits
9.4 The Role of Profinite Groups in Algebra and Topology
9.5 Predicates Defined by Systems of Equations on the Class of Groups
9.6 Interpretation of Systems of Equations over Groups of Factors that Describe a Smart System
9.7 P-topology
9.8 Pro-P-algebraic Systems

In item 9.1 we consider the meaning of the P-pure embeddings and several examples of P-purities in the class of all groups. In item 9.2 here we dwell briefly upon the modeling of "identical" factors with respect to the structure that act on the system. The question arises as to how all possible structures of connections between factors acting on the system can be described. We shall use the automorphism group of the group of factors that determine the system to this purpose.

After that we recall the definition and basic information about algebraically compact groups that are necessary for the study of innovative and pseudo-innovative systems. Algebraically compact groups are in some way a generalization of divisible groups in two following directions: The first line (1) is distinguished as a direct summand from the group containing it, when (2) certain conditions are imposed on how the subgroup is contained in the overgroup. If a divisible group can be defined as a group distinguished as a direct summand from any group that contains it, then an algebraically compact group is a group distinguished as a direct summand from any group that contains it as a pure subgroup. In item 9.3 we present some needed facts about direct and inverse spectra and their limits. In item 9.4 we recall the definition and basic information about profinite groups necessary for studying the formalization of innovative and pseudo-innovative systems. A topological group that can be represented as a projective limit of finite groups is said to be a profinite one. The class of profinite groups coincides with the class of compact completely disconnected groups. The concept of the profinite group has been time and again generalized; see, for example, Colin David Reid's paper about finiteness properties of profinite groups. Thus, for example, classes of pro-p-groups, where p is a prime number, pro-π-groups, where π is the set of prime numbers, pronilpotent groups, and pro-solvable groups were defined. In this item we introduce the notions of P-finite groups and pro-P-groups and construct some aspects of the analogous theory to the theory of algebraically closed abelian groups. The main theorem runs as follows.

Theorem *Let P be a predicate given on a class of groups which is closed under taking subgroups and factor-groups. Every group G, which is satisfied to predicate P can be embedded into pro-P-completion $\widehat{G_P}$ of a group G.*

In item 9.5 we change the definition of the purities on predicates, eliminating the condition that the predicate P should be closed with respect to taking the subalgebras. The main definition runs as follows.

Definition A subalgebra $\bar{B} = \langle B|\{f_\alpha^{n_\alpha}|\alpha \in \Gamma\}\rangle$ of an algebra $\bar{A} = \langle A|\{f_\alpha^{n_\alpha}|\alpha \in \Gamma\}\rangle$ is called a P-pure subalgebra of an algebra $\bar{A}$, if every homomorphism $\bar{B} \xrightarrow{\alpha} \bar{C}$ from a subalgebra $\bar{B}$ of an algebra $\bar{A}$ into an algebra $\bar{C}$ of a signature $\{f_\alpha^{n_\alpha}|\alpha \in \Gamma\}$, such that $P(\bar{C})$ is true, where predicate P is sustainable with respect to factor-algebras, can be continued up to homomorphism from algebra $\bar{A} = \langle A|\{f_\alpha^{n_\alpha}|\alpha \in \Gamma\}\rangle$ into an algebra $\bar{C} = \langle C|\{f_\alpha^{n_\alpha}|\alpha \in \Gamma\}\rangle$, that is the following diagram is commutative:

$$0 \rightarrow \bar{B} = \langle B|\{f_\alpha^{n_\alpha}|\alpha \in \Gamma\}\rangle \xrightarrow{\varphi} \bar{A} = \langle A|\{f_\alpha^{n_\alpha}|\alpha \in \Gamma\}\rangle$$

$$\alpha \searrow \qquad \swarrow \beta$$

$$\bar{C} = \langle C|\{f_\alpha^{n_\alpha}|\alpha \in \Gamma\}\rangle$$

It means that $\beta\varphi = \alpha$, where φ is an embedding $\bar{B} = \langle B|\{f_\alpha^{n_\alpha}|\alpha \in \Gamma\}\rangle$ into $\bar{A} = \langle A|\{f_\alpha^{n_\alpha}|\alpha \in \Gamma\}\rangle$, P is a predicate defined on the class of algebras of the signature $\{f_\alpha^{n_\alpha}|\alpha \in \Gamma\}$, highlights the class of subalgebras which is closed under taking factor algebras[2] φ is called a P-pure embedding. For such a definition of the P-purity there will be no duality, which is analogous to the duality described by L. Fucks. In item 9.6 we consider examples of the description of the functioning of systems with the help of systems of equations over groups of factors. In item 9.7 we consider the well-known concepts of topology following P. M. Cohn and construct some analogues of definitions and theorems formulated and proved by him. And finally in item 9.8 we examine the definition of direct and inverse spectra of algebraic systems and their limits for the sequel and generalize some results of previous section to the common case of algebraic systems.

In Chap. 10 we marked that the question of the reliability of the obtained results is of great value for any theory. This is especially important when it comes to risk-free application of the theoretical results in practice. The reliability is especially significant for the humanities relating to the development and functioning of human society, such as pedagogy, the general theory of education, e-learning, economics, finance as their distinctive features are the following:

- impossibility of repetition of the experiment and frequently to perform the only experiment with sufficient accuracy, since there is always the human factor,
- the difficulty of collecting reliable and comparable statistical data in connection with the lack of standardized procedures.

[2]The main operations of the same type of algebraic systems of the same signature will be denoted in each of the algebras in the same way.

In this chapter we continue to study smart systems and, in particular, the concept of smart university in the context of theoretical justification of the results based on the algebraic formalization of the smart systems. The practical result of this investigation is the evaluation of sustainability of ranking universities systems.

Chapter 10 consists of the following items:

10.1 Sustainability: Ranking Systems
10.2 Final Sustainability of a System
10.3 Time Structure of Algebraic Formalization
10.4 The Algorithm of Determination of the Scenarios of Development of the System S and Points and Intervals of Loss of the Sustainability of the System S
10.5 The Connection between Notions of Final Sustainability, Stationary Points, and Classical Sustainability
10.6 Practice Example. Algebraic Formalization as a Tool of Assertion the Sustainability of Ranking Systems of an Evaluation of Activities of Universities

In item 10.1 we marked that in the study of system's functioning across the time and its ability to forecast changes of system's properties the question about system sustainability is rather important. This question is especially important for the Smart System Theory. The concept of sustainability is well studied in terms of the availability of various quantitative parameters describing the dynamic behavior of the system. There were introduced such concepts as Lyapunov sustainability, Zhukovsky sustainability. We shall consider discrete systems as in previous chapters. Under the sustainability of a discrete system, we shall understand its ability to return to the equilibrium position after the end of the action of external factors as in the case of continuous-time systems. To date the classification of such concepts as an equilibrium, as a notion of stationary point there were introduced. The indices characterizing the quality of discrete systems designed to evaluate the dynamic properties of the system, manifested in transient conditions, and to determine the accuracy of the system which is characterized by errors in the steady state after the transition was introduced. Dynamic indicators of quality characterize the behavior of free components of the transition process closed control systems or processes of an autonomous system. However, convenient integrated indicators which are a synthesis of qualitative and quantitative indicators of the phenomenon under study as such are absent. We propose to use Cayley table of a group G_S of factors determining the system S to characterize the quality of dynamics of the closed associative smart system with feedback S. This makes it possible to regulate the behavior of the smart system S in some cases. In item 10.2 we concern the notion of a final sustainability of a system. The link between the final sustainability and Lyapunov sustainability is reviewed. In item 10.3 the time factor is introduced into the construction of the group of factors G_S determined the system S to have an opportunity to characterize the scenarios of development of the system S. In item 10.4 an algorithm to determine the points (intervals) of the loss of a sustainability of a system S and scenarios of functioning of a system S is constructed. Examples of a usage of parametric statistic in part of laws of distribution of discrete random variables in an annex to the scenarios of development of the system S are discussed. An algorithm to define and regulate scenarios of system's functioning where a system is defined by a group of factors G_S of order p^2 for a prime number p is built. In item 10.5 the connection between of the notion of final stability, stationary points, and the classical notion of sustainability is discussed.

The main result of this section runs as follows.

Theorem *If a system is final sustainable, then it is Lyupunov sustainable.*

In item 10.6 as an example, we consider a way of formalizing a synthesis of a system by its decomposition with the usage the technique of the theory of extensions of abelian groups. After that on this basis we examine the sustainability[3] of the ranking systems of evaluation the effectiveness of universities.

This construction, that is ranking system, can be used for building ranking systems monitoring smart universities. Herewith, the blocks of ranking systems themselves will change, because in this case one will have to evaluate and compare: systems for monitoring the results of the educational process, expert communities, active educational technologies, modules of educational resources, quality of IT technologies, system of formation of individual educational trajectories, technology to determine the personality characteristics of a student, the effectiveness of financial support for the activities of a smart university, and others. This will help to create a monitoring of the education system that tracks the quality of education better than existing ranking systems of an evaluation of activities of universities. Using both of these theorems and the tensor estimate of system's functioning considered in Chap. 6, we can construct new ranking system to monitor and to manage Smart Education System. It is also important that it will help to make Smart Education System more sustainable.

Reference

1. Bucur, I., Delianu, A.: Theory of categories. In: General Algebra, vol. 2, p. 188. (1968) [Chapter 7: Categories]

[3]Let us explain the notion of a sustainability of a system once more. The system is a sustainable one if at withdrawing it from the external effects from the state of equilibrium (rest) it returns to it after the cessation of external influences. From the point of view of an algebraic formalization, it means that there are restrictions on the number of final states of the system.

Chapter 1
The Problem of General Systems Theory's Formalization

Abstract In this chapter, three basic questions are considered:

- what is the meaning of the concept of "formalization"?
- how to build a formalization, allowing one to obtain and justify meaningful results in the General theory of systems, not limited to empirical reasons?
- how to interpret the results of General systems theory to specific spheres of human activity, such as the General theory of training, IT-technology, Economy and Finance, as the link with almost all types of modern human activities (e-learning, IT, economics and finance)?

The key point for us here is the connection between logic and algebra. Shafarevich (Basic concepts of algebra. Izhevsk Republican Printing House, Izhevsk, 1999, [1]), observed that the present period of development of sciences is characterized by the mathematization of the science. Algebra has always occupied a leading position in mathematics.

Keywords Formalization · Smart system theory · Algebraic systems

1.1 The Concept of Formalization as a Tool to Study the Phenomena, Processes and Practical Outcomes on a Theoretical Level

The development of System approach and a lot of works devoted to the results in General System's Theory brought up the question in what language these results should be expressed and how these results should be justified. System approach specifies General scientific methodology, so the justification of the results in this area should not be only empirical. More and more works of different complexity and different expressive means that offer various formal languages and approaches to describe the General System theory appear, [2].

Let us give a commonly accepted intuitive definition of an abstract system used in many of the already known attempts to formalize a concept of a system. The

N. Serdyukova and V. Serdyukov, *Algebraic Formalization of Smart Systems*,
Smart Innovation, Systems and Technologies 91,
https://doi.org/10.1007/978-3-319-77051-2_1

system is the minimum set of elements connected by a certain structure which gives this set of elements certain qualities that ensure the achievement of the system's goal.

One of the first works devoted to the systems' formalizations the book of Mesarovich and Takahara [3], in which the hub questions are considered:

- necessary and sufficient conditions for the controllability of multidimensional systems;
- the problem of minimum realization of regularities connecting the input effect on the system with the output parameters;
- necessary and sufficient conditions by Lyapunov stability for dynamic systems;
- Gödel's incompleteness theorem;
- decomposition problem;
- the problem of classification of systems based on the theory of categories.

In [3] the following definition of an abstract system S is given:

Let $\bar{V} = \{V_i | i \in I\}$, where I is the set of indices, be a set. Then a system S set on $\bar{V}$ is a proper subset S of a Cartesian product $\prod_{i \in I} V_i$:

$$S \subseteq \prod_{i \in I} V_i.$$

All the components $V_i, i \in I$, of a Cartesian product $\prod_{i \in I} V_i$ is called the objects of a system S.

M. Mesarovich and Y. Takahara mainly consider systems with two objects the input one X and the output one Y:

$$S \subseteq X \times Y.$$

They formalize the system in terms of binary relations on sets expressing the relationship between the properties of the system. It captures the essence of the system's studies aimed at clarifying the structure of the system and its interconnections. However, binary relations on abstract sets studied still not good enough to obtain substantial theorems of General Systems Theory with their help.

This raises three basic questions:

- what is the meaning of the concept of "formalization"?
- how to build a formalization, allowing one to obtain and justify meaningful results in the General theory of systems, not limited to empirical reasons?
- how to interpret the results of General systems theory to specific spheres of human activity, such as the General theory of training, IT-technology, economy and Finance, as the link with almost all types of modern human activities. (e-learning, IT, economics and finance)?

Let us consider the first of these questions, the question about the meaning of the concept of "formalization".

So, let us explain the concept of formalization[1] and after that we shall introduce the notion of an algebraic formalization of the system on this basis. The concept of formalization is in deep connection with logic and begins with the work by Gilbert [4], who has proposed a formalization of the process of logical deduction. The key point for us here is the connection between logic and algebra. Shafarevich [1], observed that the present period of development of sciences is characterized by the mathematization of the science. Algebra has always occupied a leading position in mathematics. Since the midst of 19th century the process of an algebraization of logic takes place.[2]

This process has become currently a very powerful and branched one[3]. There were come in sight such examples as the closure operator with the certain properties, the topological space, the monoid, the category, and so on. The reason for this were the continual presence of logical systems, the direction of studying logical systems in the fashion of some constructions, the development of the information society, the needs of insertional modeling, development of smart systems [5]. In order to clear up what is the formalization let us briefly examine the notion of an algebraic system.

1.2 Two Directions of Development of Logic. From Deductive Systems to A. I. Mal'tsev's Systems

There are two directions of development of logic: a deductive system and sequent calculus. The first direction, the deductive system, was determined by the work of Gilbert, and later on by the works of Gabbay [6, 7]. The second direction was determined by Gentzen's works, and in further developed by works of Wansing [8], in which the generalization of Gentzen's calculus (sequent calculus) was constructed [9]. The purpose of both of these trends—the search for a unified

[1]The formalization is the mapping of thinking results in precise terms or statements.

[2]From the preface of the editor of the English edition of the book R. Feys "Modal logic" … By May 1959 the overall plan appeared to him in the following form: Volume I. Theory. Volume II. Syntax and algebra. Volume III. Semantics and interpretation.

[3]Karpenko [2] "Firstly, that is a continual variety of logical systems; secondly, an extension of logical system as a result of its restriction; thirdly, an embedding (translation) of richer logical systems into weaker systems; fourthly, a tendency to the investigation of classes of logical systems in the fashion of certain constructions; fifthly, a completion of algebraization of logic; sixthly, the demands of computer revolution, etc. The result of logical concretization of two mathematical constructions—a closure operator introduced at the beginning of this century and a concept of category appeared at the middle of the century—was the beginning of investigations of different arrays of logical systems in a form of some lattice or category construction. Logic apparently loses the quality of science concerned with correctness of reasoning (and this is the phenomenon which expresses a crisis of modeling of truly human logic) and becomes the science about constructions that have an extremely abstract logical nature".

description of various logical systems. One of the possible tools to achieve this goal is the developing a common semantic foundation for a variety of logical systems.

The first direction. The deductive systems. An example of a deductive system is a Gilbert deductive logic statements, or propositional calculus [2], or propositional calculus Gilbert's type. A huge number of works is devoted to a detailed description of the propositional calculus [2].

Formal languages. Let us give the definition of a formal language. First of all let us dwell on a concept of the alphabet.

The concept of the alphabet. An arbitrary non-empty finite set of symbols is called an alphabet. A finite chain (including chain, containing no characters) which is recorded in a row of characters of the alphabet Σ is called a word in the alphabet Σ, and the number of characters (not necessarily distinct) in this chain is called the word length. The set of all words in the alphabet Σ is denoted by Σ^*. We shall use the following notation $\rceil, \wedge, \vee, \supset$ to indicate logical connections.

$\rceil A$ the negation A ("it is not true that A "),
$A \wedge B$ the conjunction A and B ("A and B"),
$A \vee B$ the disjunction A or B ("A or B"),
$A \supset B$ the implication with premise A and conclusion B ("from A follows B").

A propositional formula (a propositional logic formula, or simply a formula) is given by the following inductive definition:

(1) any propositional variable is a propositional formula,
(2) the symbols $\boldsymbol{F}$ and $\boldsymbol{T}$ (respectively false and true) are propositional formulas,
(3) if A and B are propositional formulas, then $\rceil A, (A \wedge B), (A \vee B), (A \supset B)$ is a propositional formula.

Let us designate the set of all propositional formulas by PL and let us call PL the language of propositional logic or propositional language.

The language semantics of propositional logic. Semantics or interpretation is the meaning of the words, concepts and judgments. The problem of semantics acquires a precise meaning in connection with the construction and the study of formal systems, in the case when the system receives an interpretation that is construed as displaying some meaningful theory or branch of science. The language semantics of propositional logic is given as follows. Propositional variables take values in the set $\{\boldsymbol{F}, \boldsymbol{T}\}$ or $\{0, 1\}$. We associate the truth value $v(p) \in \{0, 1\}$ with each propositional variable p where $v : var \to B = \{0, 1\}$ is the interpretation or model of the language PL. Truth-values of formulas in the interpretation of v is determined in the usual way [2], using the truth tables for logical connections on the basis of the inductive definition of formulas. A formula A which is true in every interpretation is called a valid formula, or an identically true formula or a tautology or a logical law. A tautology A is denoted as follows: $\models A$. A formula that is false in any interpretation is called an identically false one or a contradictory. From now on it is necessary to consider the concept of calculus and output in calculus. Calculus defines some basic properties (axioms) and rules (rules of inference), from which one can obtain new objects from the original ones and from previously obtained.

Definition Let Σ be an arbitrary alphabet, L be an arbitrary language over the alphabet Σ. It is said that a calculus (or a deductive system $\boldsymbol{C}$) is given if it is given a set of words A_x in the language L and a finite set R each element of which is n ary relation, where $n \geq 2$. Each element of the set A_x is called an axiom of calculus $\boldsymbol{C}$. Each element of the set R is called the rule of calculus $\boldsymbol{C}$. Every element of a set R is called a rule of inference of the calculus $\boldsymbol{C}$. Σ is called the alphabet of the calculus $\boldsymbol{C}$, L is called the language of the calculus $\boldsymbol{C}$. Sometimes alongside with the entry $\boldsymbol{C}$ it is used the entry $\boldsymbol{C}(L)$. Let r be an inference rule which is $n+1$ ary relation. If $\langle \propto_1, \ldots, \propto_n, \propto \rangle \in r$ then it is said that $\propto$ is obtained from $\propto_1, \ldots, \propto_n$ according to the inference rule r. Every word $\propto_i, i = 1, \ldots, n$, is called the premise, and the word $\propto$ is called the conclusion of the rule. It should be used the following entry in this case:

$$\frac{\propto_1, \ldots, \propto_n}{\propto}$$

Propositional calculus of Gilbert type. Propositional calculus of Gilbert type in the language of propositional logic PL is given by the axioms:

(1) $A \supset (B \supset A)$,
(2) $(A \supset (B \supset C)) \supset ((A \supset B) \supset (A \supset C))$,
(3) $A \supset (B \supset A \wedge B)$,
(4) $A \wedge B \supset A$,
(5) $A \wedge B \supset B$,
(6) $(A \supset C) \supset ((B \supset C) \supset (A \vee B \supset C))$,
(7) $A \supset A \vee B$,
(8) $B \supset A \vee B$,
(9) $(A \supset B) \supset ((A \supset \rceil B) \supset \rceil A)$,
(10) $\rceil\rceil A \supset A$,
(11) $\boldsymbol{F} \supset A$,
(12) $A \supset \boldsymbol{T}$,

where A, B, C are arbitrary formulas. The only rule of inference in this calculus is the rule modus ponens or separation rule:

$$\frac{A, A \supset B}{B}$$

With the help of output in this calculus one can get all the identically true (universally valid formulas) formulas and only them. The output of the word $\propto$ in the calculus $\boldsymbol{C}$ is the output $\propto_1, \ldots, \propto_n$, in which $\propto_n$ coincides with $\propto$. The output word $\propto$ in calculus $\boldsymbol{C}$ is denoted by $\vdash \propto$ or $\vdash_{\boldsymbol{C}} \propto$.

Let us now state the deduction theorem for the propositional calculus of Gilbert's type [10].

The theorem of deduction in the propositional calculus of Gilbert's type. Let Γ be an arbitrary set of formulas and A, B be arbitrary formulas. Then $\Gamma, A \vdash B$ if and only if when $\Gamma \vdash A \supset B$.

The following algorithm can be used to the problem of verification the deducibility in Gilbert's type calculus: by filed on the input formula the truth table is constructed and the answer "derivable" is given if its validity is found, in other case the response "not deducible" is given. This algorithm always terminates and produces the correct answer by Theorems on correctness and completeness of the calculus of Gilbert type. The method of calculating the output of Gilbert type of statements going way from the bottom up and is extremely difficult. To search the output of any given formula in the calculus of Gilbert type, one can use the so-called[4] British Museum's algorithm: sequentially generate all outputs until an output of a given formula be encountered. The algorithm generates a lot of unnecessary formulas and is therefore ineffective. The algorithm of the British Museum is open indefinitely if a given formula is unbearable.

The second direction. Sequent calculus. Another embodiment of the propositional calculus, simplifying the process of inference is Gentzen sequent calculus. Sequent calculus allows purposefully to seek inference of a given propositional formula using the decomposition of the initial inference of the search problem into simpler subtasks of the search. The main theorem of Gentzen shows that each inference can be reduced to a certain normal form although not unique. In such inference there were not introduce any other terms, except those contained in the final result, and therefore they need to be used to obtain this result. Gentzen sequent calculus is of important mathematical value for Computer Science, because it allows one to simplify the process of inference and not to look over through all the formulas in the process of inference.[5]

Boolean algebras

At the same time with the direction of developing logic as a deductive system being elaborated in the works of Gilbert, Gentzen, Frege, Russell, Whitehead [2, 11, 12], G. Boole, Jevons AI, CS Pierce, E. Shreder [2], there was offered a completely different approach to logic, which allows to investigate the basic properties of logical operations and produce logic in the form of structures which are known as Boolean algebras. Gradually algebraization of logic led to the emergence of the term "algebraic logic" [2, 12]. The methods of universal algebra are applied to the study of logic in algebraic logic.

Definition Non empty set $\boldsymbol{B}$ with two binary operations $\cup, \cap$ and one unary operation $\overline{\square}$ on it is called Boolean algebra if the following axioms are true:

(1) $A \cup B = B \cup A, A \cap B = B \cap A,$

---

[4]American scientists Allen Newell, Cliff Shaw and Herbert Simon called this procedure the British Museum algorithm," since "the idea seemed as insane as trying to seat the monkeys in front of typewriters in the hope that they will reproduce all the books of the British Museum".

[5]Boolean algebra is the result of the algebraic formalization of propositional logic [2].

(2) $A \cup (B \cup C) = (A \cup B) \cup C, A \cap (B \cap C) = (A \cap B) \cap C,$
(3) $(A \cap B) \cup B = B, (A \cup B) \cap B = B,$
(4) $A \cap (B \cup C) = (A \cap B) \cup (A \cap C), A \cup (B \cap C) = (A \cup B) \cap (A \cup C),$
(5) $(A \cap \bar{A}) \cup B = B, \quad (A \cup \bar{A}) \cap B = B.$

Thus Boolean algebra is a distributive lattice in which every element has a complement. Let us now recall the definition of a topological Boolean algebra.

Definition Boolean algebra $\boldsymbol{B}$ with the unary operation $\boldsymbol{I}$ on it is called a topological Boolean algebra if it satisfies the following conditions:

(1) $\boldsymbol{I}(A \cap B) = \boldsymbol{I}(A) \cap \boldsymbol{I}(B),$
(2) $\boldsymbol{I}(A) \leq A,$
(3) $\boldsymbol{I}(\boldsymbol{I}(A)) = \boldsymbol{I}(A),$
(4) $\boldsymbol{I}(V) = V,$

where V is an identity element of Boolean algebra that is $A \cup V = V, A \cap V = A; A \leq B$ if and only if $A \cup B = B$ or that is the same $A \cap B = A . A \cap B = A.$ Relation $\leq$ is an order relation on $\boldsymbol{B}$, that is $\leq$ is reflexive, symmetric, and transitive on $\boldsymbol{B}$. Algebraic logic is a good tool to determine the issue of how different logical systems to relate to, and to clarify the status of logic. Algebraic logic is a good tool to determine the issue of how different logical systems are correlated with one another, and it allows to clarify the status of logic. In [13], it is shown that the theorems of logic are comparable with the well-known algebraic theorems.

It is known [14] that Boolean algebra and topological Boolean algebra is a model for the classical propositional logic and modal logic[6] S4, respectively.

Boolean algebra, partially ordered sets and lattices

Let us recall that a binary relation $\rho \subseteq A \times A$ is called a partial order (or non-strict partial order) on A if ρ is transitive, reflexive and antisymmetric on A. We also recall the following definition from [16]. Element m from a partially ordered set (so called poset)[7] M is called an upper bound for a subset $N \subset M$, if every element from N is comparable with m and does not exceed m. If among all upper bounds of N there is the smallest one, then it is called a supremum of N. Similarly the greatest lower bound of the set N is defined and it is called an infimum. A partially ordered set M is called an upper semi lattice if every pair of its elements has a supremum and it is called a lower semi lattice if every pair of elements of M has an infimum. If any non-empty subset of a partially ordered set M has a supremum and an infimum,

[6]Modality is a category that expresses the degree of certainty of judgment [15]. Modal logic studies not only affirmation and negation, but also the so-called strong and weak points of affirmation and negation. Strengths include, for example, such as “truly necessary”, “necessary false”. Weak points include for example, “possible false”, “possible true”. Modal Logic (from the Latin the modus—a way to measure) is the logic, in which in addition to the standard logical connectives, variables and/or predicates there are modality or modal operators.

[7][17].

then partially ordered set (poset) M is called a complete lattice. Let us designate the supremum of elements a and b by $a \vee b$ and the infimum of a and b by $a \wedge b$. Then the following relations hold for all a and b of M:

(1) $a \vee a = a, \quad a \wedge a = a,$
(2) $a \vee b = b \vee a, \quad a \wedge b = b \wedge a,$
(3) $(a \vee b) \vee c = a \vee (b \vee c), \quad (a \wedge b) \wedge c = a \wedge (b \wedge c),$
(4) $a \wedge (a \vee b) = a, \quad a \vee (a \wedge b) = a,$

Now let us recall the definition of a semi lattice [1].

Definition [1] An algebra $\mathbf{A} = \langle A, +, \cdot \rangle$ with two binary operations + and $\cdot$ is called a lattices if the operations in it are connected by the following relations:

(1) $a + a = a, a \cdot a = a$
(2) $a + b = b + a, a \cdot b = b \cdot a$
(3) $(a + b) + c = a + (b + c), (a \cdot b) \cdot c = a \cdot (b \cdot c)$
(4) $a \cdot (a + b) = a, a + a \cdot b = a$

Elements 0 and 1 of a lattice $\mathbf{A} = \langle A, +, \cdot \rangle$ are called it's zero and unit respectively if for any a of A we have: $0 + a = a, 1 \cdot a = a$. A lattice $\mathbf{A}$ is called a lattice with complements if for any a of A there exists a' of A such that $a + a' = 1, \quad a \cdot a' = 0$. A lattice $\mathbf{A}$ is called a distributive lattice if it's elements satisfy the distributive law $a \cdot (b + c) = a \cdot b + a \cdot c$ and it is called modular if it's elements satisfy the modular law $a \cdot (a \cdot b + c) = a \cdot b + a \cdot c$. The binary relation ρ of a partial order on A makes it upper semi lattice in which $a + b = a \vee b$ and $a \cdot b = a$. Every Boolean algebra, considered only with respect to the operations $\cup$ and $\cap$, is a distributive lattice with complements. Back, every distributive lattice with complements is a Boolean algebra.

Two examples of algebraic logical systems. Later on two examples of a logical system have appeared: Tarski's deductive system which uses the concept of an operator of closure and Lambek's deductive system which uses the concept of Theory of categories [2].

Tarski's deductive system and closure operator of Tarski

Topological spaces and operators. A topological space is defined by four Hausdorff's axioms:

Let X be the set of elements of arbitrary nature. Family τ of it's subsets, which are called open sets is called a topological structure (topology) if the following axioms of a topological space take place:

1. Empty set $\varnothing$ belongs to τ,
2. X belongs to τ,
3. The union of any number of sets of τ belongs to τ,
4. The intersection of a finite number of sets of τ belongs to τ.

A set X with the given on it topological structure is called a topological space.

A closure operator on the set M persistent is the mapping the set of all subsets C of the set 2^M of M into itself which possess the following properties:

- monotony: if $X \subseteq Y$, then $C(X) \subseteq C(Y)$, for any $X, Y \in 2^M$,
- reflexive: for any $X \in 2^M$ is fulfilled $X \subseteq C(X)$,
- idempotency: for any $X \in 2^M$ is fulfilled $C(C(X)) = C(X)$.

Let us designate the set of all formulas constructed from propositional variables and logical operations of the language L by . Let $X \subseteq PL$. Theory TC induced by the operator C is the set of sentences of a language, which is closed with respect to an operator C of joining the consequences. Tarski's deductive system is (X) . $C(X)$ is the smallest closed subset containing $X.C(\varnothing)$ is a system of all provable or all universally valid proposals of the theory TC. Thus a logical system $\delta = \langle M, C_M \rangle$ is an ordered pair $\langle M, C_M \rangle$, where M is a domain, and C_M is an operator of joining the consequences (closure operator). Translation or immersion the logic $\delta = \langle M, C_M \rangle$ into a logic $\varepsilon = \langle K, C_K \rangle$ is a map $f : M \to K$ such that, for any $X \subseteq M$ the inclusion $f(C_M(X)) \subseteq C_K(f(X))$ takes place, that is translation is a continuous map. The topological interpretation of the modality has been given by McKinsey and Tarski [18], and the topological completeness of modal logic *S4* has been proved.

Lambek's deductive system using Theory of categories. Eilenberg and MacLane have created the Theory of categories [19], which became a new tool for developing foundations of mathematics, including mathematical logic and construction deductive systems. Here is the definition of the category.

Definition It is said that the category $\mathfrak{S}$ is given if it is given the class of objects $Ob\mathfrak{S}$ such that:

1. For every pair of objects (A, B) of $Ob\mathfrak{S}$ it is given the set of morphisms $Hom_{\mathfrak{S}}(A, B)$, for which we shall use the designations: $u : A \to B$, or $u \in Hom_{\mathfrak{S}}(A, B)$, or $A \xrightarrow{u} B$.
2. For every triples of objects (A, B, C) of $Ob\mathfrak{S}$ a map

$\mu : Hom_{\mathfrak{S}}(A, B) \times Hom_{\mathfrak{S}}(B, C) \to Hom_{\mathfrak{S}}(A, C)$ is given, the image $\mu(u, v)$ of a pair (u, v), where $u \in Hom_{\mathfrak{S}}(A, B), v \in Hom_{\mathfrak{S}}(B, C)$, is designated by $v \circ u$ or by vu and is called the composition of morphisms u and v.

The set $Hom_{\mathfrak{S}}(A, B)$ and the composition of morphisms satisfy the following axioms:

(1) The operation of compositions of morphisms is associative one: for every triples of morphisms $A \xrightarrow{u} B \xrightarrow{v} C \xrightarrow{w} D$ the equality $w(vu) = (wv)u$ is hold.

(2) For every object A of $Ob\mathfrak{S}$ there exists morphism $1_A : A \to A$, which is called an identity morphism of the object or a unit of the object A, such that $1_A\mu = u$ and $v1_A = v$ for any morphisms $B \xrightarrow{u} A, A \xrightarrow{v} C$.

(3) If the pairs $(A, B), (A', B')$ are differed from one another than the intersection of the sets $Hom_{\mathfrak{S}}(A, B)$ and $Hom_{\mathfrak{S}}(A', B')$ is empty.

In 1968 Lambek [20, 21], presented a deductive system as a category. In the Lambek's category the objects of the category are formulas $A, B, \ldots$, and the morphisms $u : A \to B$ are the proves $A \vdash B$. A deductive system is a category if and only if $1_A \mu = u, v 1_A = v$ and $w(vu) = (wv)u$. Categorical propositional logic is discussed in detail in [22], in which multi-level deductive systems are also considered. B 1984 Wojcicki [2], reviewed the categories in which objects $\delta = \langle M, C_M \rangle$ are kind of logics where M is a domain or an universe and C_M is an operator of joining the consequences (the closure operator) satisfying the conditions (1)–(3). In these categories, which are called the propositional calculus, morphisms are translations. Logics together with translations form a complete category, that is a category with products and sums. We are framed the issue about algebraization of logic quite narrowly, virtually without affecting the whole direction defined by the works of Boole, Jevons, Pearce, Schroeder, Lindebaum, Halmos, Racewa and Sikorski [2, 23, 24].

Algebraic systems of A. I. Malt'sev [16]. Let α, β be some ordinal numbers. The type τ of order (α, β) is called a pair of maps $W(\propto) \to N, W(\beta) \to N$ of the sets $W(\propto), W(\beta)$ into the set N of natural numbers where $W(\propto)$ is the set of all ordinal numbers strictly less than the ordinal number α. Type τ will be written in the form $\tau = \langle m_0, \ldots, m_\xi, \ldots; n_0, \ldots, n_\eta, \ldots \rangle$ где $\xi < \propto, \eta < \beta$.

The algebraic system of a type τ is an object $\boldsymbol{A} = \langle A, \Omega_F, \Omega_P, \rangle$, which consists of three sets: non empty set A—a carrier or the basic set of an algebraic system $\boldsymbol{A}$, the set of operations $\Omega_F = \langle F_0, \ldots, F_\xi, \ldots \rangle$ defined on the set A for every $\xi < \propto$, the set of predicates $\Omega_P = \langle P_0, \ldots, P_\eta, \ldots \rangle$, defined on the set A for every $\eta < \beta$. Putting $\Omega = \Omega_F \cup \Omega_P$, write the system $\boldsymbol{A} = \langle A, \Omega_F, \Omega_P \rangle$, more summery, that is $\boldsymbol{A} = \langle A, \Omega \rangle$.

A system $\boldsymbol{A} = \langle A, \Omega \rangle$ is called finite if it's basic set A is finite. The algebraic system $\boldsymbol{A} = \langle A, \Omega_F, \Omega_P, \rangle$ is called an algebra if $\Omega_P = \varnothing$, and it is called a model or relational system if $\Omega_F = \varnothing$.

A map of an algebraic system $\boldsymbol{A}$ into an algebraic system $\boldsymbol{B}$ is a map of a basic set A of an algebraic system $\boldsymbol{A}$ into a basic set $\boldsymbol{B}$ of a system $\boldsymbol{B}$.

An isomorphism of an algebraic system $\boldsymbol{A} = \langle A, F_0, \ldots, F_\xi, \ldots, P_0, \ldots, P_\eta, \ldots \rangle, (\xi < \propto, \eta < \beta)$ of the type τ into an algebraic system $\boldsymbol{B}$ is a map of a basic set A of a system $\boldsymbol{A}$ into a basic set $\boldsymbol{B}$ of an algebraic system $\boldsymbol{B} = \langle B, G_0, \ldots, G_\xi, \ldots, Q_0, \ldots, Q_\eta, \ldots \rangle$ of the same type τ is a one-to-one map such that it saves the main operations and the main predicates that is satisfying the following conditions:

$$\varphi(F_\xi(x_1, \ldots, x_{m_\xi})) = G_\xi(\varphi(x_1), \ldots, \varphi(x_{m_\xi}))$$

$$P_\eta(x_1, \ldots, x_{n_\eta}) \Leftrightarrow Q_\eta(\varphi(x_1), \ldots, \varphi(x_{n_\eta})),$$

for all $x_1, \ldots, x_{m_\xi}, x_{n_\eta}$ of A, for all $\xi < \propto$, and for all $\eta < \beta$.

Definition of subsystem. Non-empty subset A_1 of a basic set A of an algebraic system $\boldsymbol{A} = \langle A, \Omega \rangle$ is called closed in an algebraic system $\boldsymbol{A}$, if A_1 is closed under every basic operation F_ξ of this system that is the result of every basic operation F_ξ on elements of A_1 belongs to A_1.

Let F_ξ^*, P_η^* be operations and predicates defined on A_1, values of which on A_1 coincides accordingly with the values of the operations $F_\xi \in \Omega$ and predicates $P_\eta \in \Omega$. An algebraic system $\boldsymbol{A}_1 = \langle A_1, \Omega^* \rangle$ where $\Omega^* = \left\{ F_\xi^* | \xi < \propto \right\} \cup \{ P_\eta | \eta < \beta \}$ is called a subsystem of an algebraic system $\boldsymbol{A} = \langle A, \Omega \rangle$ and is recorded as $\boldsymbol{A}_1 = \langle A_1, \Omega \rangle$. If $\boldsymbol{A}$ is an algebra or a model then $\boldsymbol{A}_1$ is accordingly called a subalgebra or a submodel of $\boldsymbol{A}$.

The following diagram illustrates the foregoing:

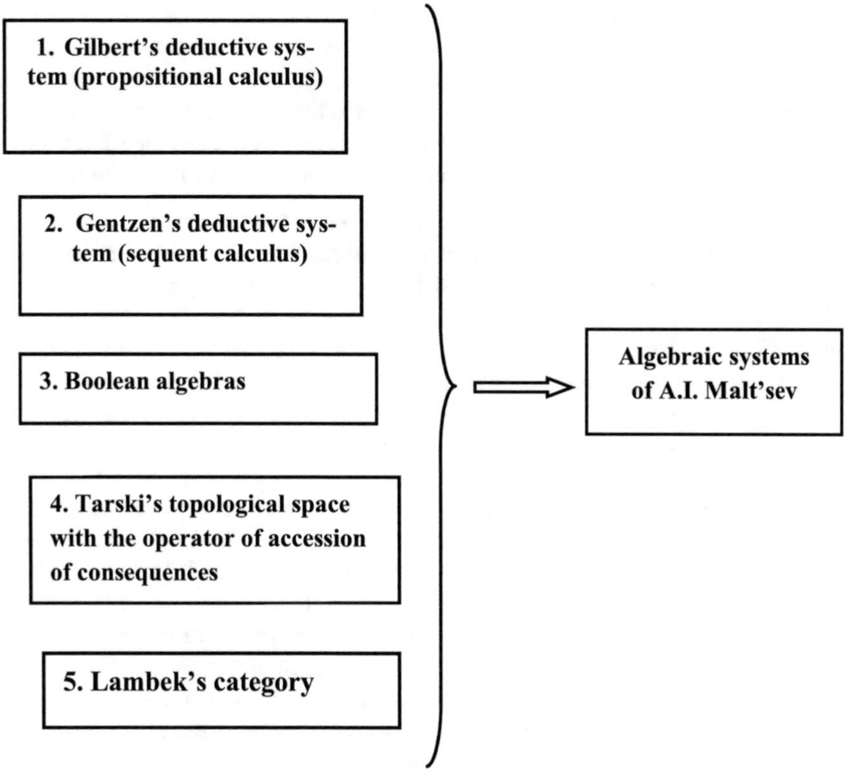

1. Gilbert's deductive system (propositional calculus)

2. Gentzen's deductive system (sequent calculus)

3. Boolean algebras

4. Tarski's topological space with the operator of accession of consequences

5. Lambek's category

⟹ **Algebraic systems of A.I. Malt'sev**

1.3 Algebraic Formalization of the General Concept of the System, Based on the Factors Determining the System

The problem of formalization of the General Systems Theory there is also due to the necessity of obtaining a priori substantive results in various spheres of human activity that are important for forecasting and planning, for example, in such areas as E-learning, Smart Systems, Finance and Economics, and so on.

Theoretical justification of empirical results
Traditionally, under the formalization we understand representation of any content area (reasoning, evidence, classification procedures, search for information, scientific theories) as a set of specific attributes for it, the use of which allows one to understand content in more meaningful way. Thus, the formalization is the mapping of thinking results in precise concepts and statements. In a developed form, these attributes appear in the form of a formal system or indicators of a calculus.

Definition Under the formalization we should understand the record or representation of any phenomena, processes and practical results designated by P which are expressed in a verbal form by the model FP of a special language $\delta = \langle M, C_M \rangle$ with its own semantics and syntax, which allows (one-to-one) display their verbal model FP in the form of symbols or constructions of this language which is called the formalization's language. We consider that it is possible to capture the essence and get new deep results in the theory of systems with the help of the theory of algebraic systems developed by Maltsev [16], and in closed associative systems—with the help of the theory of groups which is a well-established part of the theory of algebraic systems. In support of this point of view, let us remind that even A. G. Kurosh believed that the theory of groups cannot be limited only to the use in areas closest to it. We believe that this statement applies not only to the field of mathematics, but also to scientific research in general. We offer a formalization of a system based on the use of factors influencing on the system, or the factors which determined the system. P-net or P-pure embeddings on the meaning given to this term (not distort domestic relations of a system) signify such embeddings that do not distort the system's internal connections which satisfied the predicate P. We shall use them to distinguish system's relations by their nature that is to classify relations in abstract systems. The question of changing the system under the influence of factors affecting it, is the key in the study of the system behavior and its properties, so we introduce the following definition of the algebra of system factors.

Basic definitions

Definition 1.1 [25] Under the algebra of factors of a system will be understood an algebra $\bar{A} = \left\langle A | \left\{ f_\alpha^{n_\alpha} | \alpha \in \Gamma \right\} \right\rangle$ with a fundamental set of factors A and a set of operations $\left\{ f_\alpha^{n_\alpha} | \alpha \in \Gamma \right\}$ which describe connections between factors, where

n_α-is arity of operation $f_\alpha^{n_\alpha}$,
Γ-is the set of indices.

Definition 1.2 [25] A sub-algebra $\bar{B} = \langle B | \{ f_\alpha^{n_\alpha} | \alpha \in \Gamma \} \rangle$ of an algebra $\bar{A} = \langle A | \{ f_\alpha^{n_\alpha} | \alpha \in \Gamma \} \rangle$ is called P-pure in $\bar{A}$ (or an embedding φ of a sub-algebra $\bar{B}$ into an algebra $\bar{A}$ is P-pure, if (1) every homomorphism $\bar{B} \xrightarrow{\alpha} \bar{C}$ of the subalgebra $\bar{B}$ into $\bar{C}$ (where $\bar{C}$ is an algebra of the signature $\{ f_\alpha^{n_\alpha} | \alpha \in \Gamma \}$ of $\bar{A}$, and (2) $P(\bar{C})$ is true, (3) P is a predicate on the class of algebras of the signature $\{ f_\alpha^{n_\alpha} | \alpha \in \Gamma \}$ closed under taking subalgebras and factor algebras, can be continued to a homomorphism β of $\bar{A} = \langle \bar{A} \{ f_\alpha^{n_\alpha} | \alpha \in \Gamma \} \rangle$ into $\bar{C} = \langle C | \{ f_\alpha^{n_\alpha} | \alpha \in \Gamma \} \rangle$ in such a way that the following diagram is commutative:

$$
\begin{array}{ccc}
0 \to \bar{B} = \langle B | \{ f_\alpha^{n_\alpha} | \alpha \in \Gamma \} \rangle & \xrightarrow{\varphi} & \bar{A} = \langle A | \{ f_\alpha^{n_\alpha} | \alpha \in \Gamma \} \rangle \\
\alpha \searrow & & \swarrow \beta \\
& \bar{C} = \langle C | \{ f_\alpha^{n_\alpha} | \alpha \in \Gamma \} \rangle &
\end{array}
\tag{1.1}
$$

that is $\beta\varphi = \alpha$. (A note: The general operations of the same type in algebraic systems will be denoted in identical manner).

In fact, purities are the fractality of links. In this case, P-purities are the fractality of links with the property P.

Formalization expressive properties and hierarchy of formalization. Later on multi sort first-order logic, infinite logic, ω-logic, R-logic, a weak second-order logic which improve the expressive properties of formalization have appeared. Multi sort logic used, for example, in the insertion simulation. Insertion modeling is a technique of designing systems based on the theory of agents and media, applied to the verification tasks in different domains. Insertion modeling is a technique of designing systems based on the theory of agents and media, applied to the verification tasks in different domains. One of the main applications of this technology is the VRS system (Verification of Requirement Specification). VRS software package provides a means of checking the requirements for hardware and software systems by automated theorem proving and usage of symbolic deductive techniques and model validation.

Multi-sorted first order logic [26]. Two-sorted first-order logic differs from the usual first-order logic in that it contains two different kinds of variables: two-sorted algebraic system $\langle M, N, \Omega \rangle$ consists of two conventional algebraic systems with the addition of functions and predicates in their union.

ω-logic If we take a two-sorted language and consider the two-sorted algebraic system $\langle M, \boldsymbol{R}, \Omega \rangle$ with fixed system $\boldsymbol{R}$ we get $\boldsymbol{R}$-logic [26–29]. $\boldsymbol{R}$-logic is useful to study metric spaces. If $\boldsymbol{N}$ is a system of natural numbers, then $\boldsymbol{N}$-logic is called ω-logic. If R is infinite then $\boldsymbol{R}$-logic is stronger than the first order logic.

Weak second-order logic. Weak second-order logic is constructed as follows. Let L be a first order language, and x, y, z be variables of the language L. Let us construct from the language L double sorted language L^* with variables a, b, c and a symbol of adjunction $\in$. Let's extend a system $\boldsymbol{M} = \langle M, \Omega \rangle$ for the language L to the system hereditarily finite sets $\boldsymbol{HF}(\boldsymbol{M})$ over $\boldsymbol{M}$ as follows:

Let $HF_0(M) = \varnothing$,

$$HF_{n+1}(M) = \{\text{all finite subsets}\, M \cup HF_n(M)\},$$
$$HF(M) = \bigcup_n HF_n(M).$$

Let $\boldsymbol{HF}(\boldsymbol{M}) = \langle M, HF(M), \in \restriction M \cup HF(M), \Omega \rangle$. In the weak second-order logic it is allowed to use formula of L^* and to interpret the variables a, b, c by the sets of $HF(M)$. Weak second-order logic is an attempt to construct a concept of finite of logic in a natural way, it is possible to construct in it the natural numbers and to define the concept of a finite sequence. Weak second-order logic has the same effect as that of ω-logic.

Infinite logic. Logic $L_{\omega_1,\omega}$ admits a countable number of conjunctions and a countable number of disjunctions, and only a finite number of quantifiers.

The logic with new quantifiers. Mostovski proposed to introduce new quantifiers into the first-order logic. Let Q be a new symbol and let's apply the following rule of forming to it: if $\varphi(x)$ is a formula, then $Q(x)\varphi(x)$ is a formula too. For example, you can get the logic with a quantifier "there is a lot of uncountable", in which the compactness theorem holds, the logic with a quantifier "there are infinitely many" can be obtained.

And here the question arises: how to set the hierarchy on the set of formalizations?

1.4 The Hierarchy of Algebraic Formalizations

How to construct the lattice of logics? In fact it can be done using A. I. Malt'sev's algebraic systems taking into account the fact that any logic essentially is determined by some algebraic system of A. I. Malt'sev. Let's define the lattice of logics $\boldsymbol{L} = \langle L, \cup, \cap \rangle$ as a lattice of algebraic systems. Proceed as follows to construct the hierarchy of algebraic formalizations. Let's define a binary relation ρ on the set of all algebraic systems $\{\boldsymbol{A}_\propto = \langle A_\propto | \Omega_F^\propto, \quad \Omega_P^\propto \rangle, \propto \in \wedge\}$ in the following way:

Let

$$\boldsymbol{A}_\propto = \langle A_\propto | \Omega_F^\propto, \Omega_P^\propto \rangle \rho \boldsymbol{A}_\beta = \left\langle A_\beta \middle| \Omega_F^\beta, \Omega_P^\beta \right\rangle,$$

if the following embeddings are true: $A_\propto \subseteq A_\beta, \Omega_F^\propto \subseteq \Omega_F^\beta, \quad \Omega_P^\propto \subseteq \Omega_P^\beta$.

Then ρ is a reflexive, transitive and antisymmetric relation on the set of all algebraic systems. Thus the binary relation of partial order ρ is given on the set of all algebraic systems and thus ρ specifies the lattice of all algebraic systems. The set of all algebraic systems with a partial order relation ρ is a complete upper semilattice. Really, let $N = \{\boldsymbol{A}_\gamma | \gamma \in \Gamma\} \subset M$. An algebraic system with the basic set $\left\langle \bigcup_{\gamma \in \Gamma} A_\gamma \right\rangle$ of the signature $\left\langle \bigcup_{\gamma \in \Gamma} \Omega_F^\gamma, \bigcup_{\gamma \in \Gamma} \Omega_P^\gamma \right\rangle$, where by $\left\langle \bigcup_{\gamma \in \Gamma} A_\gamma \right\rangle$ we shall denote $\bigcup_{\gamma \in \Gamma} A_\gamma$, to which the results of all operations from $\bigcup_{\gamma \in \Gamma} \Omega_F^\gamma$, applied to elements of the set $\bigcup_{\gamma \in \Gamma} A_\gamma$, are added, is a supremum of the set N. So we have the complete upper semilattice of algebraic systems. An algebraic system $\left\langle \bigcap_{\gamma \in \Gamma} A_\gamma, \bigcap_{\gamma \in \Gamma} \Omega_{F\gamma}^\gamma, \bigcap_{\gamma \in \Gamma} \Omega_{P\gamma}^\gamma \right\rangle$ is an infimum of the set N, differs from $\varnothing$ if $\bigcap_{\gamma \in \Gamma} A_\gamma \neq \varnothing$. Besides it, if we add to the set of all algebraic systems empty set and allow to consider empty algebraic system with empty basic set, empty basic operations and empty basic predicates, that is if we allow to consider $\langle A, \varnothing, \varnothing \rangle$, where A can be empty set including, then we get a complete lattice of all algebraic systems. We also need the following theorem of Birkhoff: Every distributive lattice is isomorphic to some lattice of sets, [23].

1.5 Probabilistic Algebraic Formalization

Let $\boldsymbol{N} = \langle N, \vee \wedge \rangle$, where $N = \{\boldsymbol{A}_\gamma = \langle A_\gamma, \Omega_F^\gamma, \Omega_P^\gamma \rangle | \gamma \in \Gamma\}$, be a complete lattice of algebraic systems and let $\widehat{\mathrm{E}} = \left\langle \left\langle \bigcup_{\gamma \in \Gamma} A_\gamma \right\rangle, \bigcup_{\gamma \in \Gamma} \Omega_F^\gamma \bigcup_{\gamma \in \Gamma} \Omega_P^\gamma \right\rangle$. Let's construct the operation $\prime$ in the following way:

$$\boldsymbol{A}' = \left\langle \left\langle \widehat{\mathrm{E}} \backslash A_\gamma \right\rangle, \bigcup_{\gamma \in \Gamma, \propto \neq \gamma} \Omega_F^\gamma, \bigcup_{\gamma \in \Gamma, \propto \neq \gamma} \Omega_P^\gamma \right\rangle, \text{where}$$
$$\widehat{\mathrm{E}} = \left\langle \left\langle \bigcup_{\gamma \in \Gamma} A_\gamma \right\rangle, \bigcup_{\gamma \in \Gamma} \Omega_F^\gamma, \bigcup_{\gamma \in \Gamma} \Omega_P^\gamma \right\rangle, \widehat{\mathrm{O}} = \widehat{\mathrm{E}}'.$$

If for every $\boldsymbol{A}_\gamma = \langle A_\gamma, \Omega_F^\gamma, \Omega_P^\gamma \rangle \in \boldsymbol{N}$ there exists the only $\boldsymbol{A}'_\gamma$, which satisfies the conditions

$$\boldsymbol{A}_\gamma \vee \boldsymbol{A}'_\gamma = \widehat{\mathrm{E}}$$
$$\boldsymbol{A}_\gamma \wedge \boldsymbol{A}'_\gamma = \widehat{\mathrm{O}},$$

then $\boldsymbol{N} = \langle N, \vee \wedge, \prime \rangle$ is a complete distributive lattice with supplements.

Definition 1.3 Now let the set of algebraic systems N be closed under countable intersections of basic sets A_γ of algebraic systems of N and be closed under countable intersections of signatures Ω_F^γ of algebraic systems of N.

Then $\boldsymbol{N} = \langle N, \vee\wedge, \prime \rangle$ is σ-algebra and it is possible to determine the function of a probability measure $p : N \to [0; 1]$, satisfying the following conditions:

(1) $p\left(\vee_{\propto}\left(\boldsymbol{A}_{\propto} = \langle A_{\propto} | \Omega_F^{\propto}, \Omega_P^{\propto} \rangle\right)\right) = \sum_{\propto} p\left(\boldsymbol{A}_{\propto} = \langle A_{\propto} | \Omega_F^{\propto}, \Omega_P^{\propto} \rangle\right)$, where the set of indices $\{\propto | \propto \in \omega\}$ is countable, if $\langle A_{\propto} | \Omega_F^{\propto}, \Omega_P^{\propto} \rangle \wedge \left\langle A_{\beta} \middle| \Omega_F^{\beta}, \Omega_P^{\beta} \right\rangle = \widehat{\mathrm{O}}$, for any $\propto, \beta \in \omega$, such that $\propto \neq \beta$

where $\widehat{\mathrm{O}}$ is the least element of the lattice $\boldsymbol{N} = \langle N, \vee\wedge, \prime \rangle$,

(2) $p\left(\widehat{\mathrm{E}}\right) = 1$, where $\widehat{\mathrm{E}}$ is the highest element of the lattice $\boldsymbol{N} = \langle N, \vee\wedge, \prime \rangle$.

So we have a probability algebraic logic or a probabilistic algebraic formalizations in which an algebraic system $\boldsymbol{A}_{\propto} = \langle A_{\propto} | \Omega_F^{\propto}, \quad \Omega_P^{\propto} \rangle$ describes an abstract system S more reliable or more precisely than an algebraic system $\left\langle A_{\beta} \middle| \Omega_F^{\beta}, \quad \Omega_P^{\beta} \right\rangle$, if $p\left(\boldsymbol{A}_{\propto} = \langle A_{\propto} | \Omega_F^{\propto}, \quad \Omega_P^{\propto} \rangle\right) \geq p\left(\left\langle A_{\beta} \middle| \Omega_F^{\beta}, \quad \Omega_P^{\beta} \right\rangle\right)$.

Definition 1.4 Let $\boldsymbol{N} = \langle N, \vee\wedge \rangle$, where $N = \{\boldsymbol{A}_{\gamma} = \langle A_{\gamma}, \Omega_F^{\gamma}, \Omega_P^{\gamma} \rangle | \gamma \in \Gamma\}$ be a complete lattice of algebraic systems, the atoms $\Omega_N = \{\boldsymbol{A}_{\tau} | \tau \in T\}$ of this lattice is called the space of elementary algebraic formalizations.

The meaning of this definition explains the following theorem of Stone. Recall that an element a is an atom of Boolean algebra if and only if it is minimal non-zero element with respect to the induced partial order.

Stone's theorem (on the classification of finite Boolean algebras, 1936)

Every finite Boolean algebra is isomorphic to the algebra of all subsets of its atoms.

Fairly following refinement of Stone's theorem:

There exists a contravariant functor between the categories of Boolean algebras and the category of profinite spaces that is the projective limits of finite sets $F_i, i \in I$, endowed with the discrete topology.

Thus, there is a duality between the category of Boolean algebras and the category of profinite spaces.

Random formulas

Let us introduce the concept of a random formula as follows: the symbol $\boldsymbol{A} \models \varphi$ is denoted the fact that the formula φ is identically true on the algebraic system $\boldsymbol{A}$.

Definition 1.5 A formula φ of a signature Ω_{β}, where $\beta \in \Gamma, \Omega_{\beta} = \Omega_F^{\beta} \cup \Omega_P^{\beta}$, is called a random formula if for an algebraic system $\langle \boldsymbol{A}_{\gamma} | \boldsymbol{A}_{\gamma} \models \varphi$, where φ is a formula of a signature $\Omega_{\beta}, \gamma\rho\beta \rangle$, generated by algebraic systems $\boldsymbol{A}_{\gamma}$, such that $\boldsymbol{A}_{\gamma} \models \varphi$, where φ is a formula of a signature $\Omega_{\beta}, \gamma\rho\beta$ the inclusion $\langle \boldsymbol{A}_{\boldsymbol{\beta}} | \boldsymbol{A}_{\boldsymbol{\beta}} \models \varphi$, where φ is a formula of a signature $\Omega_{\beta}, \rangle \in N$ takes place.

Note that in this case $\varphi = \varphi(\boldsymbol{x})$, where $\boldsymbol{x} = (x_1, \ldots, x_n)$, or φ does not contains variables.

Definition 1.6 The probability of a random formula φ of a signature Ω_β is a probability $p(\varphi)$, defined as follows:

$$p(\varphi) = p(\langle \boldsymbol{A}_\delta | \boldsymbol{A}_\delta \models \varphi, \quad \text{where } \delta\rho\beta \rangle).$$

The value $p(\langle \boldsymbol{A}_\delta | \boldsymbol{A}_\delta \models \varphi, \quad \text{where } \delta\rho\beta \rangle)$. is defined as $\langle \boldsymbol{A}_\delta | \boldsymbol{A}_\delta \models \varphi, \quad \text{where } \delta\rho\beta \rangle \in \boldsymbol{N}$.

Let us consider the simplest properties of the function p:

(1) $p(\boldsymbol{A}_\propto \vee \boldsymbol{A}_\beta) = p(\boldsymbol{A}_\propto) + p(\boldsymbol{A}_\beta))$ if $\boldsymbol{A}_\propto \wedge \boldsymbol{A}_\beta = \widehat{\mathrm{O}}$

Definition 1.7 Let us define the distribution function $F(\varphi)$ of a random formula φ as follows:

$$F_\varphi(A_\gamma) = p(\langle \boldsymbol{A}_\delta | \boldsymbol{A}_\delta \models \varphi, \quad \text{where } \delta\rho\gamma \rangle)$$

The properties of the distribution function of the random formula

Let $\boldsymbol{A}_\propto$ be a minimum relative to ρ an algebraic system of the lattice $\boldsymbol{N}$ such that $\boldsymbol{A}_\propto \models \varphi$, and $\boldsymbol{A}_\beta$ be a maximum relative to ρ an algebraic system of the lattice $\boldsymbol{N}$ such that $\boldsymbol{A}_\beta \models \varphi$. Then $\boldsymbol{A}_\propto \rho \boldsymbol{A}_\beta$ and

$$p(\boldsymbol{A}_\propto) \leq F(\varphi) < p(\boldsymbol{A}_\beta)$$

1.6 A Series of Distribution of a Complete Countable Distributive Lattice of Algebraic Systems. A Distribution Function of a Random Function of a Lattice of Algebraic Formalizations

Definition 1.8 A function $\xi : N \to R$ defined on a complete distributive lattice of algebraic systems $N = \langle N, \vee\wedge \rangle$, where $N = \{\boldsymbol{A}_\gamma = \langle A_\gamma, \Omega_F^\gamma, \Omega_P^\gamma \rangle | \gamma \in \Gamma\}$, $\boldsymbol{A}_\beta \in N$, $\xi(\boldsymbol{A}_\beta) \in R$, and taking values in the set of all real numbers R is called a random function of a lattice of algebraic formalizations $\boldsymbol{N}$, if the complete inverse image $\xi(N) \subseteq R$ under the mapping ξ is an element of N, that is if the complete inverse image $\xi^{-1}(\xi(N)) \in N$.

A series of distribution of a complete countable distributive lattice of algebraic systems $\boldsymbol{N} = \langle N, \vee\wedge \rangle$, where $N = \{\boldsymbol{A}_i = \langle A_i, \Omega_F^i, \Omega_P^i \rangle | i = 1, 2, \ldots.\}$, $\boldsymbol{A}_i \in N, i = 1, 2, \ldots.$, is a table

Algebraic formalization	A_1	A_2	…	A_k	…
Probability	p_1	p_2	…	p_k	…

for which $\sum_{i=1}^{\infty} p_i = 1$, if the lattice $\boldsymbol{N}$ is infinite. Let us note that the top line of the table is ordered by ρ.

A series of distribution of a complete finite distributive lattice of algebraic systems $\boldsymbol{N} = \langle N, \vee\wedge \rangle$, where $N = \{\boldsymbol{A}_i = \langle A_i, \Omega_F^i, \Omega_P^i \rangle | i = 1, 2, \ldots k\}, \boldsymbol{A}_i \in N, i = 1, 2, \ldots k$, is a table

Algebraic formalization	A_1	A_2	…	A_k
Probability	p_1	p_2	…	p_k

for which $\sum_{i=1}^{k} p_i = 1$. The top line of the table is ordered by ρ.

Definition 1.9 A distribution function of a random function $\xi : N \to R$ of the lattice function of algebraic formalizations $\boldsymbol{N} = \langle N, \vee\wedge \rangle$, where $N = \{\boldsymbol{A}_\gamma = \langle A_\gamma, \Omega_F^\gamma, \Omega_P^\gamma \rangle | \gamma \in \Gamma\}, \boldsymbol{A}_\beta \in N, \quad \xi(\boldsymbol{A}_\beta) \in R$ is a function $F(x)$ such that its value in the point $x = \boldsymbol{A}_\gamma$ is equal to the probability that a random function $\xi : N \to R$ of the lattice function of algebraic formalizations $\langle N, \vee\wedge \rangle$ takes a value less than x.

Probabilistic algebraic formalization will be used in Chap. 10 to consider different scenarios of development of the system, because it randomly changes the algebraic formalization.

Let us now consider the practical examples of the use of hierarchies of algebraic formalizations in the field of intelligent (optimal) courses (smart learning course) and in the field of Abelian groups.

1.7 Examples of Usage of Hierarchies of Algebraic Formalizations

Here are examples of usage of formalized hierarchies.

1. An example of a smart learning course is the formalization of the course of higher algebra and number theory, proposed in [30].
2. In the Theory of abelian groups there is a well-known problem of describing abelian torsion-free groups of finite rank. It was solved independently by A. G. Kurosh, A. I. Maltsev and Derry in 1937. The solution they proposed used the language of matrix theory. A. Korner, solving the same problem introduced

the concept of quasi isomorphism and described torsion-free abelian groups of finite rank up to quasi isomorphism. L. Y. Kulikov and A. A. Fomin to solve the same problem used the p-adic numbers.

References

1. Shafarevich, I.R.: Basic Concepts of Algebra. Izhevsk Republican Printing House, Izhevsk (1999). (in Russian)
2. Karpenko, A.C.: Logic at the turn of the millennium. Log Investig **7**, 7–60 (2000). (in Russian)
3. Mesarovich, M., Takahara, Y.: General System Theory: Mathematical Foundations. Academic Press, New York (1975)
4. Gilbert, D., Ackerman, V.: Fundamentals of Theoretical Logic. Moscow, State Publishing House of Foreign Literature (1947). (in Russian)
5. Nikitchenko, N.S., Timofeev, V.G.: On the application of composition-nominative logics in insertion modeling. Uspekhi M **6**, 57–63 (2012). (in Russian)
6. Gabbay, D.M.: Labelled Deductive Systems, vol. 1. Clarendon Press, Oxford (1996)
7. Gabbay, D.M.: What Is a Logical System?. Clarendon Press, Oxford (1994)
8. Wansing, H.: Displaying Modal Logic. Kluwer Academic Publishers, Dordrecht (1998)
9. Vereshchagin, N.K., Shen, A.V.: Languages and Calculus, Lectures on Mathematical Logic and Theory of Algorithms. Part 2. Language and Calculus. 4 th ed. (Rev). MCNMO, Moscow (2012). (in Russian)
10. Gerasimov, A.S.: Course of Mathematical Logic and Computability Theory. 3rd ed. Publishing House "LEMA", SPb (2011). (in Russian)
11. Gentzen, G.: Investigations of Logical Inferences, the Mathematical Theory of Logical Inference. Moscow, pp. 9–74 (1967). (in Russian)
12. Whitehead, A., Russell, B.: Principia Mathematica. Cambridge University Press, England (1910–1913)
13. Halmos, P., Givant, S.: Logic as Algebra, Dolciani Mathematical Expositions, No. 21. The Mathematical Association of America, Washington (1998)
14. Khomich, V.I.: About the isomorphic embeddability of a finite Boolean and topological Boolean algebras. Logic Journal of the Study **12**, 287–292 (2005)
15. Ushakov, D.N. (ed): Dictionary of the Russian Language Four Volumes. State Publishing House of Foreign and National Dictionaries. (1935–1940). Second edition was published in 1947–1948. (in Russian)
16. Malt'sev, A.I.: Algebraic Systems. Nauka, Moscow (1970). (in Russian)
17. Garg, V.K.: Lattice Theory with Applications, Department of Electrical and Computer Engineering, University of Texas at Austin, Austin, TX 78712-1084 (2004). LNCS Homepage, https://pdfs.semanticscholar.org/acf1/00cdaac0c8c332aca7e53f82c8d447cbab56.pdf
18. Petrenko, V.F., Suprun, A.P.: Purposeful Systems. Proceedings of ISA RASB Evolution and the Subject Aspect of Systemology **62**(1), 5–27 (2012). (in Russian)
19. Eilenberg, S., Mac Lane, S.: General theory of natural equivalences. Transactions of the American Mathematical Society **58**, 231–294 (1945)
20. Lambek, J.: Deductive systems and categories I. Mathematical Systems Theory **2**(4), 278–318 (1968)
21. Lambek, J.: On the unity of algebra and logic. In: Categorial Algebra and Its Applications. Springer, Berlin (1988)
22. Vasyukov, V.L.: Categorical Logics. ANO Institute of Logics, Moscow (2005). (in Russian)

23. Rasiowa, H., Sikorski R.: The Mathematics of Metamathematics, Polska Akademia Nauk, Monografie Matematyczne, Tom 41, Panstwowe Wydawnictwo Naukowe, Warsawa (1963)
24. Sikorski, R.: Boolean Algebras. Springer, Berlin (1964)
25. Serdyukova, N.A., Serdyukov, V.I.: The new scheme of a formalization of an expert system in teaching . ICEE/ICIT 2014 Proceedings, paper 032, Riga (2014)
26. Barwise, J. (ed): Handbook of Mathematical Logic, in four parts, Part D, Proof Theory and Constructive Mathematics. Noth-Holland Publishing Company, Amsterdam (1977)
27. Barwise, J. (ed): Handbook of Mathematical Logic, in four parts, Part A, Model Theory. Noth-Holland Publishing Company, Amsterdam (1977)
28. Barwise, J. (ed): Handbook of Mathematical Logic, in four parts, Part B, Set Theory. Noth-Holland Publishing Company, Amsterdam (1977)
29. Barwise, J. (ed): Handbook of Mathematical Logic, in four parts, Part C, Recursion Theory. Noth-Holland Publishing Company, Amsterdam (1977)
30. Kulikov, L.Y.: Algebra and Number Theory. Higher School, Moscow (1979). (in Russian)

Chapter 2
The Performance of a System by Using an Algebraic System of Factors Determining the System. *P*-Properties of a System

Abstract In Serdyukova (Algebra & Logic 30:432–456, 1991, [1]), on the basis of a study Yu. L. Ershov's works about profinite groups, a method has been proposed for isolating and studying purities or pure embeddings in a special class of algebraic systems that is the class of groups, which allowed to generalize the known results of the theory of purities of abelian groups to the case of arbitrary non-abelian groups (purities by predicates). In Serdyukova (Optimization of tax system of Russia, parts I and II, 2002, [3]), Serdyukova and Serdyukov (The new scheme of a formalization of an expert system in teaching, 2014, [4]), Serdyukova et al. on the basis of system approach. Springer, Berlin, pp 371–380, 2015 [5]) and Serdyukova and Serdyukov (Modeling, simulations and optimization based on algebraic formalization of the system, pp 576–582, 2015, [6]) a method for modeling the final states of the system and determining the number of final states using the technique of group theory has been developed. An important question when studying the properties of a system and the process of its functioning, and in particular when studying the properties of a smart system and its functioning, is the question of how to determine that a system or a smart system ceases to satisfy some property P or some complex of properties Π . To answer this question, we introduce the notion of a partial probability measure on the set of unary predicates defined on the class of groups and closed with respect to taking subgroups and factor-groups.

Keywords Predicate · Algebraic system of factors determining a system
P-property

N. Serdyukova and V. Serdyukov, *Algebraic Formalization of Smart Systems*,
Smart Innovation, Systems and Technologies 91,
https://doi.org/10.1007/978-3-319-77051-2_2

2.1 Factors Determining the System

2.1.1 Static and Dynamic Predicates

Predicate P, i.e. function with a set of values from two elements $\{0, 1\}$ or $\{false, true\}$, that is, in fact, a condition that determines some property of the set separates the static properties of the system if it does not depend on time or on changing other external in relation to the system of factors. For example, for a class of algebraic systems, or for a class of all groups, or for a class of all abelian groups, the property of purity is a static one. Predicate P can single out dynamic properties of the system if it depends upon time or upon the changing of other factors external to the system. For example, if we consider financial systems and knowledge systems about them, then predicates that highlight financial sustainability, the legal sector of the economy, etc. are dynamic ones, that is, they depend on time, on the changing internal conditions of the functioning of society, and so on. For learning systems, predicates that highlight levels of learning complexity which, in turn, depend upon the degree of development of society, and so on, are dynamic. Predicates, in contrast to numerical indicators, allows us to characterize the studied properties in a single integrated complex of both numerical indicators and synchronized with them links, and in the dynamic, if they are dynamic predicates, and in static if they are static predicates. First we need to define the notion of a dynamic predicate.

Definition 2.1 A predicate P is said to be dynamic if it is represented in the form $P(X, t)$, where t is the time or some varying external factor with respect to the system X. For t, both continuous and discrete scales can be defined.

Definition 2.2 Subalgebra $\overline{B} = \langle B | \{f_\alpha^{n_\alpha} | \alpha \in \Gamma\}\rangle$ of an algebra $\overline{A} = \langle A | \{f_\alpha^{n_\alpha} | \alpha \in \Gamma\}\rangle$ is called a P-pure subalgebra of an algebra $\overline{A}$, if any homomorphism $\overline{B} \xrightarrow{\alpha} \overline{C}$ of a subalgebra $\overline{B}$ of an algebra $\overline{A}$ into an algebra $\overline{C}$ of the signature $\{f_\alpha^{n_\alpha} | \alpha \in \Gamma\}$, and such that $P(\overline{C})$ is true, where P is closed under taking subalgebras and factor-algebras,[1] can be extended up to the homomorphism of an algebra $\overline{A} = \langle A | \{f_\alpha^{n_\alpha} | \alpha \in \Gamma\}\rangle$ into an algebra $\overline{C} = \langle C | \{f_\alpha^{n_\alpha} | \alpha \in \Gamma\}\rangle$ in such a way that diagram (2.1) would be commutative:

[1]The similar main operations in the algebraic systems of the same signature will be denoted in the same way in each of the algebras.

$$0 \overset{\square}{\rightarrow} \bar{B} = \langle B|\{f_\alpha^{n_\alpha}|\alpha \in \Gamma\}\rangle \overset{\varphi}{\rightarrow} \bar{A} = \langle A|\{f_\alpha^{n_\alpha}|\alpha \in \Gamma\}\rangle$$

$$\alpha \searrow \qquad \swarrow \beta \tag{2.1}$$

$$\bar{C} = \langle C|\{f_\alpha^{n_\alpha}|\alpha \in \Gamma\}\rangle$$

that is $\beta\varphi = \alpha$, where $P(t)$ is a predicate given on the class of algebras of the signature $\{f_\alpha^{n_\alpha}|\alpha \in \Gamma\}$. Besides it predicate P should be satisfied the following conditions:

1. $P(A,t), A_1 \leq A \overset{\square}{\Rightarrow} P(A_1,t)$
2. $P(A,t), t_1 \leq t \overset{\square}{\Rightarrow} P(A,t_1)$

The proposed approach makes it possible to determine the equilibrium points of the system, the state of homeostasis[2] (equilibrium) of the system and the heterostasis[3] of the system, the transition to a hysteresis state and the number of different hysteresis states of the system, and also the homeorhesis[4] of the system. To describe the maximum number of possible states of system's hysteresis, it is necessary to analyze the concepts of synergetics and emergence, their similarity and difference. For such smart systems as knowledge systems, for examples, these concepts are important, since they show the necessary moments for changing and correcting the knowledge system, and, consequently, the moments when curricula and training courses should be changed.

2.2 The Scheme of the Dynamic Predicates' Functioning in Models that Are Groups

We, as before, shall describe the system S using the group of factors G_S, which determine the system S. To describe the change in the properties of the system during its operation and the possible change or adjustment of its target, we shall use dynamic unary predicates defined on the class of all groups. To characterize the

---

[2]Homeostasis is the ability of an open system to maintain the constancy of its internal state through coordinated reactions aimed at maintaining a dynamic equilibrium. The endeavor of the system to reproduce itself, restore the lost balance, overcome the resistance of the external environment.

[3]The essential difference between homeostasis and heterostasis is that the former maintains a normal stable state with the help of internal properties of the system, and the second—with the help of external factors affecting the system.

[4]Homeorhesis—the term suggested by Conrad Hal Waddington in 1957, means maintaining consistency in developing systems.

functioning of dynamic predicates in models which are groups, we want to define a partial probability measure on the class of all unary predicates defined on the class of groups and closed with respect to taking subgroups and factor-groups.

To this end, we recall the axiomatic definition of probability proposed by A. N. Kolmogorov.

The general definition of a probability measure looks as follows.

Let Ω be a non empty set, and $P(\Omega)$ be a set of all subsets of the set Ω. Let's consider a subalgebra $\langle \sum, \cup, \cap, \backslash, ' \rangle$ of an algebra $\langle P(\Omega), \cup, \cap, \backslash, ' \rangle$ where $P(\Omega)$ is a set of all subsets of the set Ω. So the main set $\sum$ of this algebra is closed under all main operations $\cup, \cap, '$: for any $A, B \in \sum$ the following inclusions : $A \cup B \in \sum$, $A \cap B \in \sum$, $A \backslash B \in \sum$, $A' \in \sum$ are true. Elements of the set Ω are called elementary outcomes. Elements of the algebra $\langle \sum, \cup, \cap, \backslash, ' \rangle$, or that is the same, elements of the main set $\sum$ of this algebra are called elementary events. Each of algebras $\langle \sum, \cup, \cap, \backslash, ' \rangle$ is called an algebra of elementary events or an algebra of subsets. Let $\langle \sum, \cup, \cap, \backslash, ' \rangle$ be closed under countable intersections that is if $A_i \in \sum$, where $i \in N$, N is a countable set, then $\bigcap_{i \in N} A_i \in \sum$.

Definition 2.3 An algebra of elementary events is called σ-algebra if it is closed under countable intersections.

Definition 2.4 Let every set A, belonging to some σ-algebra $\langle \sum, \cup, \cap, \backslash, ' \rangle$ of elementary events Ω, put into accordance a real number $p(A)$ such that $0 \leq p(A) \leq 1$, that is a function $p : \sum \to [0, 1]$ is given. Function p is called a probability measure or simply probability if the following conditions or axioms take place:

(1) Countable additivity's condition: the measure of the union of a countable number of pairwise disjoint events is equal to the sum of their measures, that is $p\left(\bigcup_{i \in N} A_i\right) = \sum_{i \in N} p(A_i)$, if $A_i \cap A_j = \varnothing$ for $i, j \in N, i \neq j$.
(2) Normalization condition: the probability of a true event is 1, i.e. $p(\Omega) = 1$.

Let us consider the scheme of the functioning of the dynamic predicate $P = P_t$, defined on a certain class of groups. Suppose, for example, that

$t = t_1$ $P_t(A)$ is true if and only if A is an abelian group,

$t = t_2$ $P_t(A)$ is true if and only if A is a metabelian group,

……………………………………………

$t = t_n$ $P_t(A)$ is true if and only if A is a nilpotent group of a class n.

Let $t \in T \neq \varnothing$. Let's consider the algebra of predicates $\langle \{P_t | t \in T\}, \vee, \wedge, \rceil, P_{\varnothing}, P_U \rangle$, where

$P_{\varnothing}(A)$ is false for any group A,

$P_U(A)$ is true for every group A,

$(P_{t_1} \vee P_{t_2})(A)$ is true if and only if when $P_{t_1}(A)$ is true or $P_{t_2}(A)$ is true,

$(P_{t_1} \wedge P_{t_2})(A)$ is true if and only if $P_{t_1}(A)$ and $P_{t_2}(A)$ are true,

$\rceil P_t(A)$ is true if and only if $P_t(A)$ is false. If the predicate $P = P_t$ is defined on some class of groups which is closed under taking subgroups and factor-groups, then an algebra $\left\langle \{P_t | t \in T\}, \vee, \wedge, \rceil, P_{\varnothing}, P_U \right\rangle$ is a Boolean algebra in some cases. For example it is so, if P outlines the class of finite groups. Predicate $\rceil P$ will be always false on this class of groups. If P outlines the class of all abelian groups then the predicate $\rceil P$ will be false on this class of groups. An algebra of unary predicates $\left\langle \{P_t | t \in T\}, \vee, \wedge, \rceil, P_{\varnothing}, P_U \right\rangle$ which are defined on the class of all groups is a Boolean algebra, but not all predicates herewith from $\{P_t | t \in T\}$ are closed under taking subgroups and factor-groups. An algebra $\langle \{P_t | t \in T\}, \vee, \wedge \rangle$ is a lattice. An algebra $\left\langle \{P_t | t \in T\}, \vee, \wedge, P_{\varnothing}, P_U \right\rangle$ is a distributive lattice with null and unit. The lattice of predicates allows one to order system P-properties.

If we do not require that each predicate P from $\{P_t | t \in T\}$ should be closed with respect to taking subgroups and factor-groups, then the algebra $\langle \{P_t | t \in T\}, \vee, \wedge, \rceil, P_{\varnothing}, P_U \rangle$ will be a Boolean algebra. So an algebra $\langle \{P_t | t \in T\}, \vee, \wedge, \rceil, P_{\varnothing}, P_U \rangle$ is a Boolean algebra if its main set $\{P_t | t \in T\}$ is closed under main operations.

Now let Ω be a set of all unary predicates on the class of all groups and $P(\Omega)$ be a set of all subsets of the set Ω. Let's consider a subalgebra $\langle \sum, \cup, \cap, \backslash, ' \rangle$ of an algebra $\langle P(\Omega), \cup, \cap, \backslash, ' \rangle$ of all subsets of the set Ω, that is the main set $\sum$ of this subalgebra is closed under all main operations $\cup, \cap, '$: for any $A, B \in \sum$ the following inclusions $A \cup B \in \sum$, $A \cap B \in \sum$, $A \backslash B \in \sum$, $A' \in \sum$ are true. Each of algebras $\langle \sum, \cup, \cap, \backslash, ' \rangle$ is called an algebra of elementary event or an algebra of subsets. Let $\langle \sum, \cup, \cap, \backslash, ' \rangle$ be closed under taking countable intersections, that is if $A_i \in \sum$, where $i \in N$, N is a countable set, then $\bigcap_{i \in N} A_i \in \sum$. An elementary event in this case is a complex of conditions imposed on the group, in our case on the groups of factors G_S, which determines the system S . The following corollary is true.

Corollary 2.6 *The algebra* $\left\langle \{P_t | t \in T\}, \vee, \wedge, \rceil, P_{\varnothing}, P_U \right\rangle$ *of unary predicates which are defined on the class of all groups is a* σ*-algebra.*

Definition 2.7 An ordered pair $\langle \Omega, \sum \rangle$, where $\langle \sum, \cup, \cap, \backslash, ' \rangle$ is an algebra of elementary events and $\sum \subseteq P(\Omega)$, where Ω is a set of all unary predicates on the class of all groups, is called a measurable space of regulation or the space of restrictions of the system S.

Definition 2.8 Let every set A, which belongs to some σ-algebra $\langle \sum, \cup, \cap, \backslash, ' \rangle$ of elementary events Ω, put in accordance a real number $p(A)$ such that, the condition $0 \leq p(A) \leq 1$ is true, that is a function $p : \sum \rightarrow [0, 1]$ is given. The function p is called a probability measure or simply probability if the following conditions or axioms are true:

(1) Countable additivity's condition: the measure of the union of a countable number of pairwise disjoint events is equal to the sum of their measures, that is $p\left(\bigcup_{i\in N} A_i) = \sum_{i\in N} p(A_i\right)$, if $A_i \cap A_j = \varnothing$ for $i,j \in N, i \neq j$.

(2) Normalization condition: the probability of a true event is 1, that is

$$p(\Omega) = 1.$$

Using the corollary, we introduce now the probability measure $\boldsymbol{p}$ on the σ-algebra of all unary predicates $\langle \{P_t | t \in T\}, \vee, \wedge, \rceil, P_{\varnothing}, P_U \rangle$ which are defined on the class of all groups.

Let Π be a set of all unary predicates given on the class of all groups and closed under taking subgroups and factor-groups, that is $\Pi = \{P | P(A) \& B \preccurlyeq A \Rightarrow P(B); P(A) \& C \trianglelefteq A \Rightarrow P(A/C)\}$.

We put the function $\boldsymbol{\mu_p}$ to be equal to the restriction of the probability measure $\boldsymbol{p}$ by the set Π.

$$\boldsymbol{\mu_p} = \boldsymbol{p} \upharpoonright \Pi.$$

Definition 2.9 We shall say that $\boldsymbol{\mu_p}$ defines a restriction of a probability measure, which we shall call a partial probability measure on the distributive lattice of unary predicates defined on the class of all groups and closed under taking subgroups and factor-groups $\langle \Pi, \vee, \wedge, P_{\varnothing}, P_U \rangle$.

The question arises as how to understand that the system has ceased to satisfy a certain set of properties $\mathcal{P} = \{P_i | i \in I\}$ and how to determine the moment when it ceases to satisfy the complex of properties $\mathcal{P} = \{P_i | i \in I\}$.

We shall use parametric statistics in the part of the distribution laws of discrete random variables in the application to the realization of the complex of properties of the system S.

Definition 2.10 Let's remind that a scalar function $\xi : \Omega \to R$, given on the space of elementary outcomes Ω is called a random variable if for any $x \in R$ the set $\{\omega \in \Omega | \xi(\omega) < x\}$ is an event, that is $\{\omega \in \Omega | \xi(\omega) < x\} \in \sum$.

Let $\chi_{\mathcal{P}}$ be a characteristic function that describes the fulfillment of a set of properties on the group of factors G_S, which define the system S, that is:

$$\chi_{\mathcal{P}}(t) = \begin{cases} 1, & \text{if at time } t \text{ the complex of conditions } \mathcal{P} = \{P_i | i \in I\} \text{is fullfiled on } G_S, \\ 0, & \text{if at time } t \text{ the complex of conditions } \mathcal{P} = \{P_i | i \in I\} \text{is not fullfiled on } G_S \end{cases}$$

Similarly, a characteristic function can be defined for each predicate $P_i \in \mathcal{P}$:

$$\chi_{P_i}(t) = \begin{cases} 1, & \text{if at time } t \text{ a condition } P_i \text{ is fullfiled on } G_S, \\ 0, & \text{if at time } t \text{ a condition } P_i \text{ is not fullfiled on } G_S \end{cases}$$

Binomial distribution for the realization of a complex of properties $\mathcal{P} = \{P_i|i \in I\}$ of a system S in n trials with the probability of successful realization of a complex of properties $\mathcal{P} = \{P_i|i \in I\}$ of a system S, equal to p.

Let $\mathcal{P} = \{P_i|i \in I\}$ be a complex of properties of a system S, which are given by unary predicates on the class of all groups that are closed under taking subgroups and factor-groups. Let's consider the binomial distribution on the algebra of unary predicates $\langle\{P_t|t \in T\}, \vee, \wedge, \rceil, P_\varnothing, P_U\rangle$ defined on the class of all groups for the realization by a system S a complex of properties $\mathcal{P} = \{P_i|i \in I\}$ with the probability of successful realization equal to $\boldsymbol{p}$.

The value of a random variable ξ is equal to the number of successful implementations by the system S a complex of properties $\mathcal{P} = \{P_i|i \in I\}$ in n trials with the probability of successful realization of the complex equal to $\boldsymbol{p}$.

ξ	0	1	…	l	…	n
Probability	$(1-\boldsymbol{p})^n$	$n\boldsymbol{p}(1-\boldsymbol{p})^{n-1}$	…	$C_n^l\boldsymbol{p}^l(1-\boldsymbol{p})^{n-l}$	…	$\boldsymbol{p}^n$

The sense of the binomial distribution for modeling the realization by the system S the complex of properties $\mathcal{P} = \{P_i|i \in I\}$ runs as follows: the probability that at least one property from the complex of properties $\mathcal{P} = \{P_i|i \in I\}$ is not realized in one test is equal to $\boldsymbol{q} = 1 - \boldsymbol{p}$.

Chebyshev's inequality gives the following estimate, called the rule of three sigma in application to the binomial law:

The probability

$$P\left(n\boldsymbol{p} - 3\sqrt{n\boldsymbol{p}(1-\boldsymbol{p})} \le \xi \le n\boldsymbol{p} + 3\sqrt{n\boldsymbol{p}(1-\boldsymbol{p})}\right)$$
$$= 1 - P\left(|\xi - n\boldsymbol{p}| > 3\sqrt{n\boldsymbol{p}(1-\boldsymbol{p})}\right) \ge 8/9 \approx 0,9$$

Whereas consider the restriction $\boldsymbol{\mu}_p = \boldsymbol{p}\restriction\Pi$, we obtain

Partial binomial distribution for the realization of a complex of properties $\mathcal{P} = \{P_i|i \in I\}$ of a system S in n trials with the probability of successful realization of a complex of properties $\mathcal{P} = \{P_i|i \in I\}$ of a system S, equal to p.

The value of the random variable ξ is equal to the number of successful realizations of the complex $\mathcal{P} = \{P_i|i \in I\}$ of the system S in n trials with the probability of successfully realizing the complex of properties $\mathcal{P} = \{P_i|i \in I\}$ of the system S equal to $\boldsymbol{p}$,

ξ	0	1	…	l	…	n
μ_p	$(1-\boldsymbol{p})^n$ // or undefined	$n\boldsymbol{p}(1-\boldsymbol{p})^{n-1}$ // or undefined	…	$C_n^l\boldsymbol{p}^l(1-\boldsymbol{p})^{n-l}$ // or undefined	…	$\boldsymbol{p}^n$ // or undefined

This construction makes sense in the case when $\boldsymbol{\mu}_p$ is defined on the elementary event $\mathcal{P} = \{P_i | i \in I\}$.

If in the distributive lattice $\langle\{P_t | t \in T\}, \vee, \wedge, \rceil, P_{\varnothing}, P_U\rangle$ one chooses the sublattice $\langle\{P_t' | t \in T\}, \vee, \wedge, \rceil, P_{\varnothing}, P_U\rangle$, in such a way that its main set $\{P_t' | t \in T\}$ is contained in the complex of properties $\mathcal{P} = \{P_i | i \in I\}$ and is a σ-algebra, for example, is a finite sublattice, then $\boldsymbol{\mu}_p$ is in an interval between two probability measures.

Let's make two important observations.

Remark 1 If the set of properties $\mathcal{P} = \{P_i | i \in I\}$ of system S changes then the system can break the development scenario G_S.

Remark 2 If the development scenario G_S of system S changes, the system can break its complex of properties

$$\mathcal{P} = \{P_i | i \in I\}.$$

In this connection, the question arises about the recognition of the properties of the system, the scenario of its development, and the regulation of the properties of the system and scenarios for its development. The question of recognition of scenarios for the development of the system was considered in Chap. 10 (Sect. 10.4).

2.3 Cycles in the System's Development and Functioning

Let the system S at the points of time $\{t_{\propto} | \propto \in \Lambda\} \subseteq R$ at the points of time possess a complex of properties $\mathcal{P}(t_{\propto}) = \{P_i(t_{\propto}) | i \in I\}$. If the relation $(\forall \propto \in \Lambda)(\forall i \in I_1 \subseteq I)(P_i(t_{\propto}) = P_i)$ takes place, then in the system S there is a partial recurrence of development on the time interval $\Lambda \subseteq R$, that is, a cycle.

Theorem 2.11 *If the binomial distribution law holds for the complex of properties $\mathcal{P} = \{P_i | i \in I\}$ of the system S, then, according to the rule of three sigma cycles in the development of the system S occur with a probability not less than $\frac{8}{9} \cong 0.9$.*

2.3.1 Cycle Length

The next interesting question is the question about the length of the cycle, that is the question arises as to how one can establish the length of the cycle in time with some degree of probability.

Let there exists an aggregate index $\vartheta(\mathcal{P}(t_\propto))$ for measuring the complex of properties $\mathcal{P}(t_\propto) = \{P_i(t_\propto)|i \in I\}$ of the system S. Then $min\vartheta(\mathcal{P}(t_\propto))$ characterizes the decreasing stage of the cycle, and $max\vartheta(\mathcal{P}(t_\propto))$ characterizes the upward stage of the cycle. $min\vartheta(\mathcal{P}(t_\propto))$ and $max\vartheta(\mathcal{P}(t_\propto))$ are random variables, since these are extremums of random variables, and therefore each of them has a distribution function: $F(min\vartheta(\mathcal{P}(t_\propto)))$ and $F(max\vartheta(\mathcal{P}(t_\propto)))$ respectively. If each of the random variables $min\vartheta(\mathcal{P}(t_\propto))$ and $max\vartheta(\mathcal{P}(t_\propto))$ satisfies the binomial distribution law, then by the rule of the three sigma we can calculate the time limits of the crisis with a probability not less than $\frac{8}{9}$.

2.3.2 Practical Examples of the Smart System Cyclic Functioning

Practical examples of cycles in the development of smart systems we find, first of all, in economic theory. For the first time economic cycles were, apparently, discovered by the French physicist Juglar [7]. In 1913 the Dutch economist Jacob van Gelderen developed the theory of undulating evolutionary development of the economy, in which he substantiated 50–60-year cycles. Juglar identified the economic cycles associated with the renewal of the active part of fixed capital with a length of 7–11 years. Also there are known Kitchin cycles with a length of 3–4 year, cycles of Simon Kuznets. At the present times there is a large number of works devoted to economic cycles [8].

The most interesting is represented by Kondratiev's theory of economic cycles, 1920, or the theory of long waves.

2.3.3 Kondratiev's Cycles in Economic Theory

It is believed that the time length of the Kondratiev's cycle, or the wavelength of Kondratiev's waves, is 50 years, with a possible deviation of 10 years (from 40 to 60 years). Kondratiev's cycles consist of alternating phases of relatively high and relatively low rates of economic growth. Kondratiev singled out the following four empirical regularities in the development of large cycles, or long waves:

(1) before the beginning of the upward wave of each large cycle, and sometimes at the very beginning of it there are significant changes in the conditions of the

economic life of society. Changes are expressed in technical inventions and discoveries, changes in the conditions of monetary circulation, and in the strengthening of the role of new countries in world economic life. Changes occur constantly, but unevenly and most intensely expressed before the onset of the upward waves of large cycles and at their onset.

(2) periods of upward waves of large cycles, as a rule, are much richer in large social shocks and upheavals in the life of society (revolutions, wars) than periods of descending waves.

(3) the descending waves of large cycles are accompanied by a prolonged depression of agriculture.

(4) large cycles of economic conjuncture are revealed in the same unified process of the dynamics of economic development, in which the average cycles are identified with their phases of recovery, crisis and depression.

For the period after the industrial revolution, the following Kondratiev's waves usually stand out:

- 1st cycle—from 1803 to 1841–1843 (the moments of the minimum economic indicators of the world economy are marked);
- 2nd cycle—from 1844–1851 to 1890–1896;
- 3rd cycle—from 1891–1896 to 1945–1947;
- 4th cycle—from 1945–1947 to 1981–1983;
- 5th cycle—from 1981–1983 to ~2018 (forecast);
- 6th cycle—from ~2018 to ~2060 (forecast).

There are differences in the dating of "post-Kondratiev" cycles, for example, the following boundaries of the beginning and the end of "post-Kondratiev" waves are also given:

- 3rd cycle: 1890–1896–1939–1950;
- 4th cycle: 1939–1950–1984–1991;
- 5th cycle: 1984–1991–?

2.3.4 J. Schumpeter Theory of Cyclic Development

Josef Schumpeter developed a dynamic concept of the economic cycle and established a connecting link between Kondratiev's long cycles and the medium-term cycles of Juglar. The relative correctness of the alternation of the upward and downward phases of Kondratiev's long waves, each phase of 20–30 years, is determined by the nature of the group of nearby medium-term cycles [9]. During the upswing of the long wave, rapid expansion of the economy leads society to the need for change. Opportunities for changing society are lagging behind the requirements of the economy, so development goes into a downward phase, during which economic and other relations contribute to crisis-depressive phenomena.

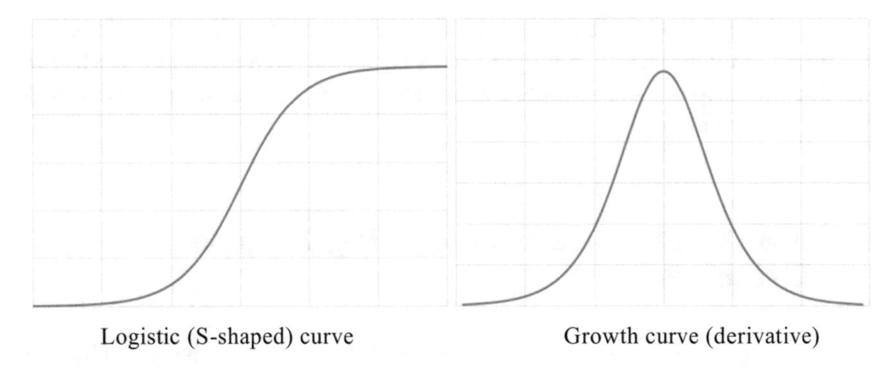

Picture 2.1 Life cycle of technological changes

At the present time in the theory of long waves the following propositions are considered to be proven [10]:

(1) Beginning with the industrial revolution of the late 18th century, quasi-cyclic oscillations with a period of about half a century are distinguished in the indicators of economic activity of the advanced countries. Unlike cyclic processes in the development of natural or technical systems, long waves do not have strict periodicity and strict recurrence of sequences of events.
(2) The dynamics of indicators, in the oscillations of which there are long waves, is asynchronous, which affects the chronology of long waves.
(3) Despite the fact that usually long waves are represented in the form of a sinusoidal reflecting the oscillations of any of the indicators, the life cycle of technological changes takes the form of a logistic curve formed from the phases of various indicators differing in growth rates (Picture 2.1).
(4) The mechanism of formation and change of long waves is multifactorial. Long waves arise as a result of a multitude of nonlinear feedbacks acting between technological, macroeconomic, institutional, social subsystems with different lags and with a high degree of uncertainty. The main issue remains the identification the logic of these connection links.
(5) The stochastic nature of the processes that take place in the economic system.

The construction of models of system dynamics began with the work of Jay Forrester, a professor at the Massachusetts Institute of Technology. Jay Forrester developed the apparatus of "system dynamics", which allows to simulate with the help of computers the development of various scenarios in the dynamics of complex systems. The device was built on the basis of the achievements of the theory of systems and computer modeling using the language of ordinary differential equations (ODE). To build the simulation models that describe the world dynamics (the so-called World-Systems), one-type ordinary differential equations of the first order are used in the form: $dy_i/dt = f_i^+ - f_i^-, i = 1, \ldots, n$. Here f_i^+ is a right-hand side of the equation, which includes all the factors that cause the growth of the variable

y_i, and f_i^- includes all factors that cause the decrease of the variable y_i . It is also assumed that these terms on the right-hand side can be represented as a product of functions that depend only on factors F_j which, in turn, are themselves functions of the main variables y_i .

2.4 Algorithm for Determining and Regulating Smart System's Properties

Under regulation the system's properties we shall understand the possibility of influencing the rate of change of processes occurring in the system. Each process that takes place in the system is determined by the properties or complex of properties that occur in the system. We shall assume that for each set of properties of the system we have a very good system of indicators that determine the change in the properties of the system. It is possible to characterize the development or functioning, or the movement of the system, if one knows some of its states and possible patterns of repetition of modifications of these states (equilibrium approach), or if one knows the patterns of self-organization and self-regulation of complex systems (synergistic approach). Within each of these approaches, the concept of system sustainability is extremely important. The sustainability property of the system is its possibility to deviate slightly from its own state under small external or internal disturbances. Let $\{P_1,\ldots,P_n\}$ be a set of unary predicates on the class of groups which describe the properties of the system S, modeled by the group of factors G_S, defining the system S. Let $\{a_{ij}|i,j=1,\ldots,n\}$ be a system of indicators which determine the implementation of the properties of the system S: an indicator a_{ij} gives a numerical estimate of the realization of the property P_i under the condition that the property $P_j, i,j=1,\ldots,n$, is fully realized when the system S is functioning. Then a_{ii} characterizes the complete realization of the property P_i by the system S. Let $M=\|a_{ij}\|\in R^{n\times n}$ be a square $n\times n$-matrix with the elements from the field of real numbers whose elements are the numerical indicators of the realization of the complex of properties $\{P_1,\ldots,P_n\}$ in the system S. Adding to this construction the time factors $\{t_\propto|\propto\in\Lambda\}\subseteq R$, we have: $M(t_\propto)=\|a_{ij}(t_\propto)\|\in R^{n\times n}$, $\propto\in\Lambda$, is a square $n\times n$-matrix with the elements from the field of real numbers whose elements are the numerical indicators of the realization of the complex of properties $\{P_1,\ldots,P_n\}$ in the system S at the time $t_\propto$, $\propto\in\Lambda$.

Definition 2.12 The determinant $|M|$ of the matrix M is called an aggregated indicator of a realization of a complex of properties $\{P_1,\ldots,P_n\}$ of the system S.

Definition 2.13 The determinant $|M(t_\propto)|$ of a matrix $M(t_\propto)=\|a_{ij}(t_\propto)\|\in R^{n\times n}$ is an aggregated indicator of a realization of a complex of properties $\{P_1(t_\propto),\ldots,P_n(t_\propto)\}$ of a system S at the time $t_\propto$, $\propto\in\Lambda$.

Let's assume that the matrix M has n-pairwise distinct eigenvalues: $\lambda_1, \ldots, \lambda_n$, that is the characteristic equation $|\lambda E - M| = 0$ of the matrix M has exactly n pairwise distinct roots $\lambda_1, \ldots, \lambda_n$. Then the aggregated parameter of realization of a complex of properties $\{P_1, \ldots, P_n\}$ of the system S equals $\prod_{i=1}^{n} \lambda_i = |M|$.

Let's assume that the matrix $M(t_\propto)$ has n-pairwise distinct eigenvalues: $\lambda_1(t_\propto), \ldots, \lambda_n(t_\propto)$, that is the characteristic equation $|\lambda E - M(t_\propto)| = 0$ of the matrix $M(t_\propto)$ has exactly n pairwise distinct roots $\lambda_1(t_\propto), \ldots, \lambda_n(t_\propto)$. Then the aggregated parameter of realization of a complex of properties $\{P_1(t_\propto), \ldots, P_n(t_\propto)\}$ of the system S at the time $t_\propto, \propto \in \Lambda$ equals $\prod_{i=1}^{n} \lambda_i(t_\propto) = |M(t_\propto)|$.

Each of $\lambda_i, i = 1, \ldots, n$ shows the rate of change of the eigenvector X of the matrix M, that is, the solution of the equation $MX = \lambda X$.

It is well known that $\sum_{i=1}^{n} \lambda_i = \sum_{i=1}^{n} a_{ii} = tr(M)$ is a trace of the matrix M.

Respectively, $\sum_{i=1}^{n} \lambda_i(t_\propto) = \sum_{i=1}^{n} a_{ii}(t_\propto) = tr(M(t_\propto))$ is a trace of the matrix $M(t_\propto)$.

Every $\lambda_i(t_\propto), i = 1, \ldots, n$ shows the rate of change of the eigenvector $X(t_\propto)$ of the matrix $M(t_\propto)$, that is, the solution of the equation $M(t_\propto)X = \lambda X$, $\propto \in \Lambda$.

Let us consider the physical meaning of the eigenvectors and eigenvalues of the matrix.

According to [11], eigenvalues are the most important feature of practically any dynamic system. The eigenvalues and eigenvectors of the matrix M appear automatically when solving a homogeneous system of linear ordinary differential equations (ODE)

$$dx/dt = A(t)x$$

of the equation $du/dt = Mu$ [12]. This is a matrix equation of the first order, linear with respect to the unknown function.

Such an equation has "purely exponential solutions" of the form $u = e^{\lambda t}x$; an eigenvalue is the rate at which an eigenvector grows or decreases x. Other solutions will be combinations of these "pure" solutions, compiled in such a way that the initial conditions are satisfied. In the general case, there are few eigenvectors and they are directed in different directions. Let us describe this in more detail, following [12].

A system

$$dx/dt = A(t)x \tag{2.2}$$

is called a normal homogeneous system of linear ODEs corresponding to the system

$$dx/dt = A(t)x + g(t) \tag{2.3}$$

If $g(t) \not\equiv 0$ in $T, T \subseteq R$ then the system will be inhomogeneous.

Let's use matrix designations:

$$x(t) = \begin{pmatrix} x_1(t) \\ \dots \\ x_n(t) \end{pmatrix}$$

$$A(t) = \begin{pmatrix} a_{11} & \dots & a_{1n} \\ \dots & \dots & \dots \\ a_{n1} & \dots & a_{nn} \end{pmatrix} \quad (2.4)$$

$$g(t) = \begin{pmatrix} g_1(t) \\ \dots \\ g_n(t) \end{pmatrix}$$

2.4.1 Vronsky's Determinant

Let there be given a system consisting of $x_1(t),\ldots, x_k(t),\ldots, x_n(t)$, $x_k(t) = (x_{1k}(t), \ldots, x_{nk}(t))^T$, $k = 1, \ldots, n$, which is defined on the interval $T \subseteq R$. The determinant

$$W(t) = \begin{pmatrix} x_{11}(t) & \dots & x_{1n}(t) \\ \dots & \dots & \dots \\ x_{n1}(t) & \dots & x_{nn}(t) \end{pmatrix}$$

is called Vronsky's determinant of the system of vector-functions $x_1(t),\ldots, x_k(t),\ldots, x_n(t)$.

Properties of Vronsky's determinant.

Theorem 2.14 *If the system of vector functions $x_k(t) = (x_{1k}(t), \ldots, x_{nk}(t))^T$, $k = 1, \ldots, n$, is lineary dependent on some interval $T \subseteq R$, then Vronsky's determinant $W(t) \equiv 0 \, \forall \, t \in T$.*

Theorem 2.15 *If the Vronsky's determinant $W(t)$ of the system of vector-functions*

$$x_k(t) = (x_{1k}(t), \ldots, x_{nk}(t))^T, k = 1, \ldots, n, \quad (2.5)$$

which are solutions of the homogeneous system of linear ODEs (2.2) *in the interval $T \subseteq R$ is equal to zero, although at one point $t_0 \in T$, then this system of vector functions is linearly dependent in the interval T.*

Corollary 2.16 *If the Vronsky's determinant $W(t)$, composed of vector-functions that are solutions of the homogeneous system of linear ODEs* (2.2) *in some interval T, is zero at one point $t_0 \in T$, that is $W(t_0) = 0$, then it is identically equal to zero in this interval.*

Definition 2.17 A linearly independent system in the interval $T \subseteq R$ of n vector functions $x_k(t) = (x_{1k}(t), \ldots, x_{nk}(t))^T$, $k = 1, \ldots, n$, each of which is a solution of a homogeneous system of n linear ODEs (2.2), in this interval is called the fundamental system of solutions for (2.2) in this interval.

Theorem 2.18 Fundamental systems of solutions exist.

2.4.2 *Theorems on the Structure of the General Solution of a Homogeneous System of Linear ODEs*

Theorem 2.19 If the vector function

$$x_k(t) = (x_{1k}(t), \ldots, x_{nk}(t))^T, \quad k = 1, \ldots, n,$$

defined on the interval $T \subseteq R$, form on this interval a fundamental system of solutions of a homogeneous system of linear ODEs

$$dx/dt = A(t)x, \tag{2.2}$$

then the general solution of this system has the form

$$x(t) = \sum_{k=1}^{n} C_k x_k(t) \tag{2.6}$$

with some constant coefficients C_k, $k = 1, \ldots, n$.
The set of solutions of the homogeneous system (2.2) forms an n-dimensional vector space, and any fundamental system of solutions serves as a basis in this space.

Definition 2.20 The equation

$$\begin{vmatrix} a_{11} - \lambda & \ldots & a_{1n} \\ \ldots & \ldots & \ldots \\ a_{n1} & \ldots & a_{nn} - \lambda \end{vmatrix} = 0 \tag{2.7}$$

is called the characteristic equation of a system of linear ordinary differential equations of ODE with constant coefficients

$$dx/dt = Ax \tag{2.8}$$

The following two cases are possible:

(1) All the roots $\lambda_1,\ldots,\lambda_n$ of the characteristic equation (2.7) are real and distinct. The general solution of a homogeneous system of linear ODEs with variable coefficients in the particular case of system (2.8) with constant coefficients can be written in the form

$$x(t) = \sum_{j=1}^{n} C_j x_j(t) = \sum_{j=1}^{n} C_j a_j e^{\lambda_j t}, \tag{2.9}$$

where C_j, $j = 1,\ldots,n$, arbitrary constants, or in the coordinate form

$$\left(x_1(t) = \sum_{j=1}^{n} C_j a_{1j} e^{\lambda_j t}, \ldots, x_i(t) = \sum_{j=1}^{n} C_j a_{ij} e^{\lambda_j t}, \ldots, x_n(t) = \sum_{j=1}^{n} C_j a_{nj} e^{\lambda_j t}\right) \tag{2.10}$$

(2) Suppose that there are complex roots among the simple roots of the characteristic equation. Such roots are divided into pairs of complex conjugate roots $\lambda = a + bi, \bar{\lambda} = a - bi, i^2 = -1$, since the characteristic equation (2.7) is an n-th power equation with real coefficients. For $\lambda = a + bi$ let's consider the complex solution of a system of homogeneous linear equations:

$$\begin{aligned} (a_{11} - \lambda)\alpha_1 + a_{12}\alpha_2 + \cdots + \alpha_{1n}\alpha_n &= 0 \\ \cdots \\ a_{n1}\alpha_1 + a_{n2} \propto_2 + \cdots + (a_{nn} - \lambda) \propto_n &= 0 \end{aligned} \tag{2.11}$$

and the complex-valued vector function $x(t) = \propto e^{\lambda t} = (p + iq)e^{\lambda t}$, from which Euler's formula yields:

$$\begin{aligned} x(t) &= \propto e^{\lambda t} = (p + iq)e^{\lambda t} = (p + iq)e^{at}(\cos bt + i \sin bt) \\ &= e^{at}(p \cos bt - q \sin bt) + ie^{at}(q \cos bt + p \sin bt), \end{aligned}$$

or $x(t) = u(t) + iv(t)$, where

$$u(t) = e^{at}(p \cos bt - q \sin bt), v(t) = e^{at}(q \cos bt + p \sin bt).$$

The vector-function $x(t) = \propto e^{\lambda t}$ satisfies the ODE system (2.8). The complex-valued function $\bar{x}(t) = \bar{\propto} e^{\bar{\lambda} t}$ will satisfy the ODE system (2.8). Therefore, the vector functions $u(t)$ and $v(t)$ are real and imaginary parts of the complex-valued function $x(t) = \propto e^{\lambda t}$ are solutions of the ODE system (2.8). Besides it, $u(t)$ and $v(t)$ are linearly independent over R.

(3) The characteristic equation (2.7) has multiple roots. Let's number each multiple root of the characteristic equation (2.7) as many times as the linearly

independent eigenvectors of the matrix A of the ODE system (2.8) corresponds to it. We have only $s \leq n$ roots $\lambda_1, \ldots, \lambda_s$, to each of which there corresponds an eigenvector of the matrix A. The following theorem holds under these conditions.

Theorem 2.21 *There exists a system of n linearly independent vectors* $l_k^{(j_k)}, j_k = 1, \ldots, q_k, k = 1, \ldots, s$, *satisfying relations*

$$
\begin{aligned}
Al_k^{(1)} &= \lambda_k l_k^{(1)};\\
Al_k^{(2)} &= \lambda_k l_k^{(2)} + l_k^{(1)};\\
&\ldots.\\
Al_k^{(q_k)} &= \lambda_k l_k^{(q_k)} + l_k^{(q_k-1)},
\end{aligned}
$$

where the sum q_k, *corresponding to the same* λ_k, *equals to the multiplicity of this root.*

Theorem 2.22 *To each root* λ_k *corresponds* q_k, *where* q_k *is the multiplicity of the root* λ_k, *solutions of the system (2.8) of the form*

$$
\begin{aligned}
x_k^{(1)} &= l_k^{(1)} e^{\lambda_k t}\\
x_k^{(2)} &= (l_k^{(2)} + t l_k^{(1)}) e^{\lambda_k t}\\
&\ldots.\\
x_k^{(q_k)} &= \left(l_k^{(q_k)} + t l_k^{(q_k-1)} + \ldots + \frac{t^{q_k-1} l_k^{(1)}}{(q_k-1)!} \right) e^{\lambda_k t}.
\end{aligned}
$$

The immediate consequence of is the following theorem.

Theorem 2.23 *If the determinant of the matrix* $M(t_\propto)$ *is nonzero in the time interval* $T = \{t_\propto | \propto \in \Lambda\} \subseteq R$, *then it is possible to control the processes determined by the complex of conditions* $\{P_1(t_\propto), \ldots, P_n(t_\propto)\}$ *of the system S, and occurring in the system S in the time interval* $T = \{t_\propto | \propto \in \Lambda\} \subseteq R$.

In fact, in this case, according to all above, at each time instant of the interval $T = \{t_\propto | \propto \in \Lambda\} \subseteq R$ there is known a set of n pairwise distinct eigenvalues of the matrix $M(t_\propto)$, that determine the rates of change of vectors of eigenvectors of matrix $M(t_\propto)$, which determine the direction of development of the system S. By influencing the speed of these processes, for example, in educational or economic systems, we can within certain limits regulate the processes occurring in the system, i.e., to affect their speed, quality, intensity, etc., in accordance with the purpose of regulation.

2.4.3 Algorithm of Determining the Possibility of Regulating the Properties of the System S

Let us now turn to the algorithm for determining and regulating the properties of the system S.

Step 1. Solve the characteristic equation

$$|\lambda E - M| = 0 \tag{2.12}$$

Suppose that it has n pairwise distinct non-zero solutions: $\lambda_1, \ldots, \lambda_n$. These solutions show, respectively, the rates of change of the eigenvectors of the matrix M.
Step 2. We introduce the partial order relation $\lesssim$ on the set of properties $\{P_1, \ldots, P_n\}$. We assume that $P_i \lesssim P_j$, if the property P_j prevails over the property P_i, that is if the property P_j is executed in the system S then the property P_i is also satisfied in the system S:Let's designate it as follows:

$$(S \models P_j) \Rightarrow (S \models P_i)$$

Step 3. We choose the maximal elements of the set $\{P_1, \ldots, P_n\}$ with respect to the partial order $\lesssim$. Suppose that this is $P_1, \ldots, P_k$.
Step 4. We regulate the velocities $\lambda_1, \ldots, \lambda_k$ of the processes of the system S, determined, respectively, by the properties $P_1, \ldots, P_k$ of the system S .

References

1. Serdyukova, N.A.: On generalizations of purities. Algebra & Logic **30**(4), 432–456 (1991)
2. Ershov, YuL: Profinite groups. Algebra and Logic **19**(5), 552–565 (1980)
3. Serdyukova, N.A.: Optimization of Tax System of Russia, Parts I and II. Budget and Treasury Academy, Rostov State Economic University, Moscow (2002). (in Russian)
4. Serdyukova, N.A., Serdyukov, V.I.: The new scheme of a formalization of an expert system in teaching. ICEE/ICIT 2014 Proceedings, paper 032, Riga (2014)
5. Serdyukova, N.A., Serdyukov, V.I., Slepov, V.A.: Formalization of knowledge systems on the basis of system approach. In: SEEL2015, Smart Education and Smart e-Learning, Smart Innovation, Systems and Technologies, vol. 41, pp. 371–380. Springer, Berlin (2015)
6. Serdyukova, N.A., Serdyukov, V.I., Modeling, simulations and optimization based on algebraic formalization of the system. In: 19th International Conference on Engineering Education July 20–24, 2015, Zagreb, Zadar (Croatia), New Technologies and Innovation in Education for Global Business, Proceedings, pp. 576–582, ICEE2015, Zagreb (2015)
7. Juglar, C.: Des Crises commerciales and their return periodic in France, England and the United States, Paris (1862)
8. Sadovnichiy, V.A., Akayev, A.A., Korotayev, A.V., Malkov, SYu.: Modeling and Forecasting World Dynamics. ISPI RAS, Moscow (2012). (in Russian)
9. Kondratiev, N.D.: Problems of economic dynamics. Economics, Moscow (1989). (in Russian)

10. Akaev, A.A., Korotaev, A.V., Malinetskiy, G.G., Malkov, S.Yu. (ed): Modelirovanie and Prediction of Global, Regional and National Development. Book House "LIBROM", Moscow (2012). (in Russian)
11. Streng, G.: Linear Algebra and Its Applications. Mir, Moscow (1980). (in Russian)
12. Zarubin, V.S., Krishchenko, A.P. (ed): Complex of textbooks from 21 issues, XVI. Theory of Probability, Moscow, Izdatelstvo MGTU im. N.E. Bauman (2004). (in Russian)

Chapter 3
The Simulation of the System with the Help of Finite Group of Factors Determining the System. *P*-Properties of the System. Cayley Tables and Their Role in Modeling Associative Closed System with Feedback

Abstract Communications among the systems factors are characterized by the interactions between the systems factors. In our case these are a binary operation of composition of the factors and an unary operation $\square^{-1}$, which characterized the feedback of the system. So Cayley table [1], of the group $\boldsymbol{G}_S$ of factors describing the system S plays an important role in describing the system's connections. In this chapter it is from this position we shall begin to consider the question of the sustainability of the system which is defined in fact by the internal structure of its connections, robust and interchangeability of structural resources. There after we shall continue to study the sustainability of the system in the Chap. 10 from the position of the usage of numerical characteristics associated with Cayley table of the group of factors determining the system. Some useful ideas can be find in [2, 3].

Keywords Group of factors determining the system · Cayley table
Sustainability

3.1 *P*-Properties of the Smart System. Sustainability of Smart Systems

Under the sustainability of the system we shall mean the system's ability to save its current state upon the influence of external and internal influences. If the system S is able to save a certain state $S = S_e$ in time moments $t \in T$, where T is a continuous or a discrete set, that is if $S(t) = S_e,\ t \in TS(t) = S_e,\ t \in T$ under the influence of external and internal factors, then the state S_e is called an equilibrium state of the system S. Any algebraic relations in the group $\boldsymbol{G}_S$ will define communications in the

N. Serdyukova and V. Serdyukov, *Algebraic Formalization of Smart Systems*, Smart Innovation, Systems and Technologies 91,
https://doi.org/10.1007/978-3-319-77051-2_3

system S and so the system's sustainability to some extent. If for example we have a system of equations

$$\bigwedge_{i \in I} w_i(x_1, \ldots, x_{n_i}) = e, \tag{3.1}$$

then we can consider that it represent some connections between the elements of set of its solutions in $\boldsymbol{G}_S$.

Exactly from this position to study the property of sustainability of system the notions of quotient-rigid and quotient-flexible systems are introduced. In the Chap. 3 with the same position we propose the following partial classification of the property of the sustainability of the system, which complemented the concept of P-quasi-sustainable system:

- the compensational sustainability or the factors' sustainability of the system S for the interchangeably factors a_i and a_j for the quotient—flexible systems,
- the sustainability with regard to the system's target of the system S which is described by the finite group of factors $\boldsymbol{G}_S$,
- the quasi—sustainability with regard to predicate which includes as a special case the sustainability with regard to the system's target,
- the final sustainability of the system which we shall consider in the Chap. 10.

In this chapter we also show that we can restrict the study of the infinite system by the usage of finite sets, namely finite sets of factors that determine the system.

In the Chap. 10 we shall establish a connection between the concepts of the final sustainability of the system and Lyapunov sustainability [4–6], of the system.

One of the most important issues of the System Theory solutions for which the theory of finite groups can be used is a question about the sustainability of a system. Sustainability is a primary quality of any system. In the absence of this quality a system cannot exist. Sustainability brings together the various properties: resistance to external factors, sustainability, reliability, etc. The sustainability of a system as a property is displayed in the system's responses onto a perturbation of various kinds. Nevertheless we can say that at present time the term "sustainability" is not sufficiently clear from its content. In a more rigorous analysis of dynamic continuous systems the notion of sustainability has multiple interpretations, namely:

- this is a property of a system to deviate little from some state during little perturbations,
- this is a property of a system to this is a property of a system in a process of a motion to remain in a restricted area of a phase space. Under the phase space one should understand a system of coordinates, on axes of which the values that characterize the dynamic system are displayed,
- this is a property of a system in a process of a motion arbitrary late to return arbitrary close to its initial state in a phase space,

- this is a property of a system to preserve some features of the phase portrait under a small perturbation of the law of motion of a system. The phase portrait is a set of phase trajectories along which the system may move in phase space from different starting points state,
- this is a property of continuity, which must have a dependence on the state of the object or the flow of the process from a specific parameter.

New idea of a system sustainability is associated with the last interpretation. The development of this idea led to the fact that in the late 1960s—early 1970s René Thom and Christopher Zeeman introduced the terms "catastrophe" and "catastrophe theory" [7]. In 1972 René Thom published a book in which the foundations of the theory of catastrophes were outlined. Catastrophe—it is a sudden jump in the status change, a catastrophe means cutting quality change in the object with a smooth change of the quantitative parameters on which it depends. One of the main tasks of the theory of catastrophes is to obtain so-called normal form of the object—a differential equation or a map in the vicinity of "point of catastrophe" and a classification of objects built on the basis of this. René Thom introduced seven generalized structures bifurcation geometries, which correspond to the standard form of expansions in Taylor series. Then V. I. Arnold proposed a classification of accidents, use a deep connection with the theory of Lie groups [8, 9].

The question of the sustainability of a system is closely related to questions of an emergence and a decay of the system. This question is one of the first questions in the study of the sustainability of a system. Prigogine [10], gives a formula which explains the creation and functioning of a system:

$$\ldots \textbf{CHAOS} \rightarrow \textbf{SYSTEM} \rightarrow \textbf{CHAOS} \rightarrow \textbf{SYSTEM} \ldots$$

We understand chaos as a nonempty set on which no operations and predicates are defined during the transition to models of algebraic formalization of systems.

Let S be a closed associative system and $\boldsymbol{G_S} = \langle G_S, ^\circ, \square^{-1}, e \rangle$ be a group of factors which represents the system S. Let G_S be finite and consists of n elements, $|G_S| = n$. Let $G_{1S}, G_{2S}, \ldots, G_{mS}$ be all pairwise nonisomorphic groups of n elements. Let $T_{1S}, T_{2S}, \ldots, T_{mS}$ be Cayley tables for groups $G_{1S}, G_{2S}, \ldots, G_{mS}$ respectively. The following question arises: what will be happened to the system S if there will be changes in any group of factors G_{lS}, $1 \leq l \leq m$, for example:

- any factor or multiple factors would not be functioning,
- feedback would not be working,
- the property of a system to associative one would be disturbed,
- neutral factor will cease to function.

Answers to these questions help to analyze the qualitative changes occurring in the system.

A partial answer to this question can be obtained by well-known theorems of finite groups theory namely, the following statements are true.

Theorem 3.1 *Upon termination of the functioning of one of the factors from the group* $\boldsymbol{G_S}$ *of order* $|G_S| = n$ *representing the system S that is equivalent to the removal of this factor from the group of factors the system S loses the property of closeness.*

The proof is obtained directly from Lagrange theorem, as the number of factors continue to operate in the system *S*, is $n - 1$ and $n - 1$ does not divide *n*.

Theorem 3.2 *Upon termination of the functioning of m factors where m does not divide n, from the group* $\boldsymbol{G_S}$ *of order* $|G_S| = n$ *representing the system S that is equivalent to the removal of these factors from the group of factors the system S loses the property of closeness.*

Proof In this case $n - m$ does not divide *n*, and according to Lagrange theorem there is no a subgroup of order $n - m$ in $\boldsymbol{G_S}$, so the system *S* loses the property of closeness.

Theorem 3.3 *Upon termination of the functioning of one of the factors a or* a^{-1} *from the group* $\boldsymbol{G_S}$ *representing the system S, the system S loses the property of feedback on this factor.*

This can be interpreted in another way: the signature of the system changes that is the group signature $\langle *, \Box^{-1}, e \rangle$ changes for semigroup with unit signature $\langle *, e \rangle$, and one get semigroup $\boldsymbol{Smg_S}$ which represents the system *S* instead of a group of factors $\boldsymbol{G_S}$, which represents the system *S*.

Theorem 3.4 *A system S with the group of representing factors* $\boldsymbol{G_S}$ *retains the closure property after the cessation of the functioning factors* $\{a_i | i \in I\}$ *if and only if* $\langle G_S \backslash \{a_i | i \in I\}, ^\circ, \Box^{-1} \rangle$ *is a group.*

Let's consider the simplest examples to illustrate these statements.

3.2 Example. Smart Systems Modeling by a Group of Four Elements

There are exactly two pairwise groups of four elements: cyclic group of four elements

$$Z_4 = \langle F_4 \| a_1^4 = e,\ a_1^2 = a_2,\ a_1^3 = a_3 \rangle$$

Fig. 3.1 Cyclic group of order 4

*	e	a_1	a_2	a_3
e	e	a_1	a_2	a_3
a_1	a_1	a_2	a_3	e
a_2	a_2	a_3	e	a_1
a_3	a_3	e	a_1	a_2

Fig. 3.2 The Cartesian product of two cyclic groups of the 2nd order (i.e. a quad group)

*	e	a_1	a_2	a_3
e	e	a_1	a_2	a_3
a_1	a_1	e	a_3	a_2
a_2	a_2	a_3	e	a_1
a_3	a_3	a_2	a_1	e

and a quad group of four elements:

$$Z_2^2 = \langle F_4 \| a_1^2 = \mathrm{e}, a_2^2 = \mathrm{e},\ a_3^2 = \mathrm{e} \rangle,$$

which are represented by the following Cayley tables (Figs. 3.1 and 3.2):

Let's consider what will be happened with the system S if some of its factors would cease to function.

Let's designate:

p_0—the probability that neutral factor e ceased to function,
p_1—the probability that factor a_1 ceased to function,
p_2—the probability that factor a_2 ceased to function,
p_3—the probability that factor a_3 ceased to function.

Let's suppose that in the system S which is modeled by a group of factors $\mathbf{Z_4}$ factor a_1 ceased to function.

This can be illustrated as follows (Fig. 3.3):

Cayley table shows that in this case the system S with the probability p_1 loses the closure property because, for example $a_2 \cdot a_3 = a_1$. Nevertheless the remaining set of factors $\{e, a_2, a_3\}$ has a partial feedback by a factor a_2.

Let's suppose now that in the system S which is modeled by a group of factors $\mathbf{Z_4}$ factor a_2 ceased to function.

Fig. 3.3 Factor a_1 ceased to function

*	e	a_1	a_2	a_3
e	e	a_1	a_2	a_3
a_1	a_1	a_2	a_3	e
a_2	a_2	a_3	e	a_1
a_3	a_3	e	a_1	a_2

Fig. 3.4 Factor a_2 ceased to function

*	e	a_1	a_2	a_3
e	e	a_1	a_2	a_3
a_1	a_1	a_2	a_3	e
a_2	a_2	a_3	e	a_1
a_3	a_3	e	a_1	a_2

Fig. 3.5 Factor a_3 ceased to function

*	e	a_1	a_2	a_3
e	e	a_1	a_2	a_3
a_1	a_1	a_2	a_3	e
a_2	a_2	a_3	e	a_1
a_3	a_3	e	a_1	a_2

This can be illustrated as follows (Fig. 3.4):

Cayley table shows that in this case the system S with the probability p_2 loses the closure property because, for example $a_1 \cdot a_1 = a_2$.

The feedback property on the remaining factors remains.

Let's suppose now that in the system S which is modeled by a group of factors $\mathbf{Z}_4$ factor a_3 ceased to function (Fig. 3.5).

Cayley table shows that in this case the system S with the probability p_3 loses the closure property because, for example $a_1 \cdot a_2 = a_3$.

The feedback property on the remaining factors remains.

Fig. 3.6 Factors a_1 and a_3 ceased to function

*	e	a_1	a_2	a_3
e	e	a_1	a_2	a_3
a_1	a_1	a_2	a_3	e
a_2	a_2	a_3	e	a_1
a_3	a_3	e	a_1	a_2

Let's suppose now that in the system S which is modeled by a group of factors $\boldsymbol{Z_4}$ factors a_1 and a_3 ceased to function.

This can be illustrated as follows (Fig. 3.6):

Cayley table shows that in this case the system S with the probability $(1 - p_1)\,(1 - p_3)$ preserves the closure property and a feedback under the condition of an independence of factors.

Another important question that arises during the study of the properties of the stability of the system is the issue about the possibility of mutual substitution of elements of the system, or the factors determining the system or the system's functions to achieve system goals. In order to outline possible solutions of this issue, let's consider firstly the question about the relationship between the factors which determined the system and elements of the system, and then, on this basis, the issue of compensational properties of the system.

3.3 Relationship Between Factors Determining a System and Elements of a System

In the examples we have considered the representation of a system S by a group of factors $\boldsymbol{G_S}$ where group $\boldsymbol{G_S}$ was finite. During the study system's properties directly the following question arises: how one link the factors which determine the system and system's elements?

To solve this problem we shall use the following procedure.

Let system S be consist of the following elements: $S = \{s_\propto |\ \propto \in \Lambda\}$.

Suppose that the group's of factor determining the system S main set $G_S = \{e, a_1, \ldots, a_n\}$, and herewith the elements of a system S, defines each factor that is mutually relevant to each factor are set off: $a_i \leftrightarrow S_i = \{s_{\propto_i} |\ \propto_i \in \Lambda_i,\ i = 1, \ldots, n\} \neq \emptyset$.

Note that each subset $S_i = \{s_{\propto_i} |\ \propto_i \in \Lambda_i,\ i = 1, \ldots, n\}$ of the set S one-to-one corresponds the subset $\{\Lambda_i | i = 1, \ldots, n\}$ of the set Λ.

The following cases are possible:

(1) $\{\Lambda_i | i = 1, \ldots, n\}$ is a splitting of a set Λ. This means that $\bigcup_{i=\overline{1,n}} \Lambda_i = \Lambda$, and $\Lambda_i \cap \Lambda_j = \emptyset$ for any $i, j \in \{1, \ldots, n\}$ such that $i \neq j$.
(2) $\{\Lambda_i | i = 1, \ldots, n\}$ is not a splitting of a set Λ. By virtue of the definition of the group of factors which determine the system one can assume without loss of generality that $\bigcup_{i=\overline{1,n}} \Lambda_i = \Lambda$. So there exist $i, j \in \{1, \ldots, n\}$ in this case such that $\Lambda_i \cap \Lambda_j \neq \emptyset$. In this case we construct the grinding of the set $\{S_i | \mathrm{i} = 1, \ldots, \mathrm{n}\}$ up to the splitting $\{S'_\mathrm{i} | \mathrm{i} = 1, \ldots, \mathrm{n}\}$ of the set S, where $S'_\mathrm{i} = S_i \backslash \left(\bigcup_{j=1, j \neq i}^{n} (S_i \backslash S_j) \right)$, $i = 1, \ldots, n$.

If the condition (2) takes place then the intersection of clusters $S_i \cap S_j$ is called a reserve of functions f_i and f_j.

Conditions (1) and (2) lead to the following definitions.

Definition 3.5 The system S is called a quotient—rigid one if the condition (1) is true. The system S is called a quotient—flexible one by factors a_i and a_j if $\Lambda_i \cap \Lambda_j \neq \emptyset$, $i, j \in \{1, \ldots, n\}$.

The assumption of the finiteness of the group of factors G_S which determine the system S is not essential.

Thus, without loss of generality we can assume that each factor a_i, $i \in I$ from the group of factors G_S which determine a system S corresponds to a cluster of elements $S_i = \{s_{\propto_i} | \propto_i \in \Lambda_i\}$ of a system S. This correspondence is one-to-one. This implies that Theorems 3.1–3.4 can be restated as follows.

Theorem 3.6 *The system loses the closure property if one cluster* $\{s_{\propto_i} | \propto_i \in \Lambda_i\}$ *of a system S, which corresponds to a factor from the group* $\mathbf{G_S}$ *would stop its functioning.*

Theorem 3.7 *The system loses the closure property if m clusters of a system S, where m does not divide n, would stop their functioning.*

Theorem 3.8 *The system loses the closure property if one cluster* $\{s_{\propto_i} | \propto_i \in \Lambda_i\}$ *of a system S would stop its functioning.*

The advantage of such formulations of Theorems 3.1–3.8 in is that we can use only finite sets to study infinite systems.

In Chap. 10 we would continue the study of sustainability of systems from the position of usage of the numerical characteristics associated with the Cayley table.

3.4 Substitution of Functions of a System. System's Compensational Possibilities

We shall follow [2] in this section. Under the substitution (compensation) of a broken function of a system we would understand the adaptation of the system to changing conditions of its existence and a replacement as the consequence as a consequence of this broken or ineffective or not working elements of a system by relatively more efficient elements of a system. We would call such elements of a system substitutional or compensational ones.

Let system S be consist of the following elements: $S = \{s_{\propto} | \propto \in \Lambda\}$.

Suppose that the group's of factor determining the system S main set $G_S = \{e, a_1, \ldots, a_n\}$, and herewith the elements of a system S, defines each factor that is mutually relevant to each factor are set off: $a_i \leftrightarrow S_i = \{s_{\propto_i} | \propto_i \in \Lambda_i, i = 1, \ldots, n\} \neq \emptyset$. Note that each subset $\Lambda S_i = \{s_{\propto_i} | \propto_i \in \Lambda_i, \ i = 1, \ldots, n\}$ of the set S one-to-one corresponds the subset $\{\Lambda_i | i = 1, \ldots, n\}$ of the set Λ.

Let $F_S = \{f_i | i \in I\}$ be functions corresponding to the set of factors $\{a_i | i \in I\}$ which determine a system S.

As it was mentioned above we can assume without loss of generality that $\{a_i | i \in I\}$ coincides with $G_S = \{e, a_1, \ldots, a_n\}$. Let us depict it through the Diagram (3.1).

This raises the following questions:

(1) how to determine what is the factor that determines a system?
(2) how to determine what is the function that corresponds to the factor determining a system?

Answers to these questions can be given by a well-known technique used in Set theory, namely one can set both of these concepts with the help of axioms.

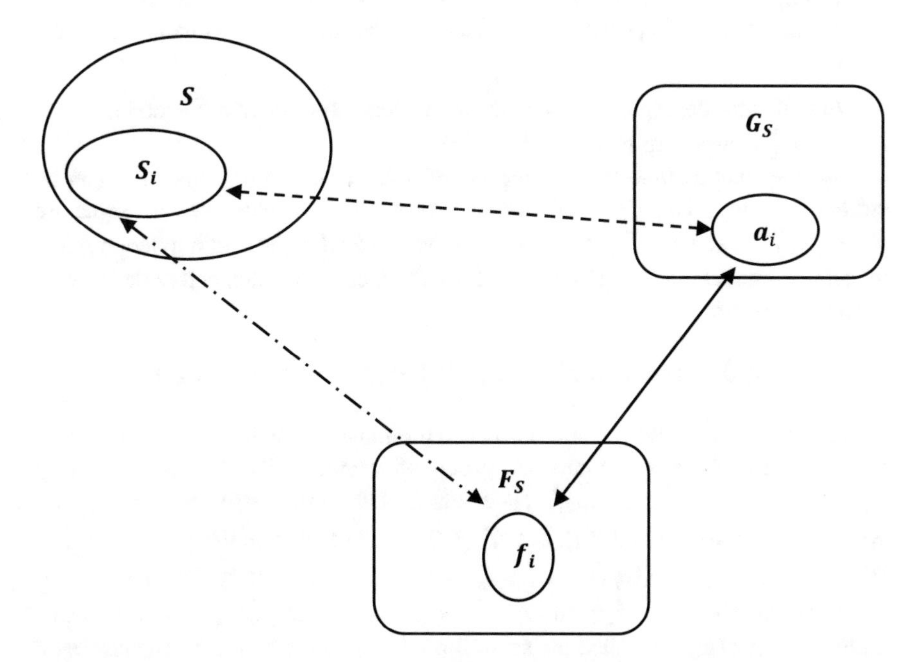

Diagram 3.1 One-to-one correspondence of factors determining the system and the elements of the system

So, we have:

Diagram (3.1) can be explained as follows if to adhere the classical scheme of the study of systems: a system—elements—the structure. We can consider the factor a_i, $i = 1,\ldots,n$, as a cloud of elements $S_i = \{s_{\propto_i} | \propto_i \in \Lambda_i,\ i = 1,\ldots,n\}$, and then to consider the structure of the factors a_i, $i = 1,\ldots,n$, or that is the same, the structure of cloud clusters $S_i = \{s_{\propto_i} | \propto_i \in \Lambda_i,\ i = 1,\ldots,n\}$.

3.5 Compensational Functions of a Quotient—Flexible Smart System

Let us consider the set of functions $F_S = \{f_i | i \in I\}$ of a system S which one-to-one correspond to the set of factors $G_S = \{e, a_1,\ldots,a_n\} = \{a_i | i \in I\}$ determining the system S, consisting of the elements $\{s_\propto | \propto \in \Lambda\}$, that is $S = \{s_\propto | \propto \in \Lambda\}$. The system S under the action of function f_i passes into a state S_i, herewith the system S converts into a model G_S^i. During this transition the following can occur:

(1) a system S will lose the closure property in case the function f_i neutralizes factor a_i, or several factors together with the factor a_i,
(2) a system S will lose the associative property,
(3) a system S will lose the feedback property,
(4) an impact of factor a_i will increase in the system S under the influence of function f_i and it is possible the impact of some several factors determining the system S.

We shall consider a partial case when after every exposure of a function f_i, $i \in I$, a model $\boldsymbol{G_S^i}$ continue to be a group.

Also we shall assume that all impacts of a function f_i onto a system S can be written in the language of narrow predicate calculus of a signature $\Omega_S = \langle \cdot, \square^{-1}, e, \{a_i | i \in I\} \rangle$. Let us denote the set of all impacts of a function f_i in a model $\boldsymbol{G_S}$ onto a system S as $Th(f_i)$ and it will be called an elementary theory of a function f_i, that is

$$Th(f_i) = \{Q_1\ldots Q_n \varphi(x_1,\ldots,x_n) | G_S^i \vDash Q_1\ldots Q_n \varphi(x_1,\ldots,x_n)\}.$$

A formula $Q_1\ldots Q_n \varphi(x_1,\ldots,x_n)$ is called a symmetric one if for any permutation $\tau \in S_n$, where S_n is a permutation group of degree n the following identity $\varphi(x_1,\ldots,x_n) \equiv \varphi(x_{\tau(1)},\ldots,x_{\tau(n)})$ takes place. Let $SymmTh(f_i)$ be a set of all symmetric formulas from $Th(f_i)$. Let $\tau Th(f_i)$ be a set of all formulas $\varphi(x_1,\ldots,x_n) \equiv \varphi(x_{\tau(1)},\ldots,x_{\tau(n)}) Q_1\ldots Q_n \varphi(x_1,\ldots,x_n) Q_1\ldots Q_n \varphi(x_1,\ldots,x_n)$ from $Th(f_i)$ such that for a transposition $\tau = (i,j)$ an equality $\varphi(x_1,\ldots,x_n) \equiv \varphi(x_{\tau(1)},\ldots,x_{\tau(n)})$ takes place. The set $\tau Th(f_i)$ is called the set of formulas symmetric on the transposition τ for a function f_i where $i \in I$.

Definition 3.9 Let $\tau = (i,j)$ be a transposition of degree n. We shall say that a factor a_i replaces a factor a_j if an equality $\varphi(x_1,\ldots,x_n) \equiv \varphi\left(x_{\tau(1)},\ldots,x_{\tau(n)}\right)$, where $Q_1\ldots Q_n\varphi(x_1,\ldots,x_n) \in Th(f_i)$ takes place.

Definition 3.10 Let us say that the system S allows functions replacement if there exists such transposition $\tau = (i,j)$ that $\tau Th(f_i) \neq \emptyset$.

The following theorem is a consequence of these definitions.

Theorem 3.11 *If the inequality $\tau Th(f_i) \neq \emptyset$ is true for a group of factors $\mathbf{G}_S$ which determines the system S, that is if a group $\mathbf{G}_S$ has a symmetric formula subset, then the system S allows functions substitution.*

Let's connect functions substitution with system sustainability.

Definition 3.12 A system S is sustainable on factors a_i and a_j if $Th(f_i) = Th\left(f_j\right)$.

Next theorem is a direct consequence of the definitions.

Theorem 3.13 *To ensure the system to be a sustainable one by the factors a_i and a_j it is necessary that it permits functions replacement.*

To associate the notions of system's sustainability on factors and final system's sustainability introduced in Chap. 10, we can use the following theorem.

Theorem 3.14 *If the system S is a final sustainable one, then the elementary theory $Th(\mathbf{G}_S)$, where $\mathbf{G}_S$ is a group of factors that determine the system S is a complete one.*

3.6 Sustainability upon the Smart System Functioning

The question about how to describe the following situation in the functioning of a system arises. Suppose that some non-empty set of elements of a system are out of order in the process of functioning of a system but however a system continues to function through its other resources and achieves its purpose. We shall offer the description of this situation for the case when the system S is described by a finite groups of factors $\mathbf{G}_S$. Let $G_S = \{e, a_1,\ldots,a_n\}$. In the Chap. 4 it would be shown that the system's target can be recorded using unary or n—ary predicate Q in the following way:

$$\bigwedge_{i\in I} P_i \Rightarrow Q$$

or

$$\bigwedge_{i\in I} P_i(S) \Rightarrow Q(A_1,\ldots,A_n),$$

where $A_1, \ldots, A_n$ are in the domain of predicate Q. Then the fact that a system S fulfills the purpose Q can be written as follows: $Q(S)$, and under the transition to a group of factors $\boldsymbol{G_S}$ which determine a system S it can be written as follows: $Q(\boldsymbol{G_S})$. In the case when the predicate Q is closed under taking subgroups and factor-groups we can speak about the sustainability by predicate Q which defines a goal of a system S.

The set of formulas

$$(\exists x_1)(\forall x)(x = x_1)$$
$$(\exists x_1)(\exists x_2)(\forall x)((x = x_1) \vee (x = x_2))$$
$$\ldots$$
$$(\exists x_1)(\exists x_2)\ldots(\exists x_k)(\forall x)((x = x_1) \vee (x = x_2) \vee \ldots \vee (x = x_k)),$$
$$\ldots$$
$$(\exists x_1)(\exists x_2)\ldots(\exists x_{n+1})(\forall x)((x = x_1) \vee (x = x_2) \vee \ldots \vee (x = x_{n+1})),$$

or that is the same we can define the formula

$$\varphi \equiv \bigvee_{k \in N} (\exists x_1)(\exists x_2)\ldots(\exists x_k)(\forall x)((x = x_1) \vee (x = x_2) \vee \ldots \vee (x = x_k))$$

which helps to write system's property in case when the system is determined by the subgroup $\boldsymbol{H}$ of k or countable number of elements of a group $\boldsymbol{G_S}$ to achieve the goal Q:

$$Q(H), \quad H \vDash (\exists x_1)(\exists x_2)\ldots(\exists x_k)(\forall x)((x = x_1) \vee (x = x_2) \vee \ldots \vee (x = x_k).$$

If such a subgroup $\boldsymbol{H}$ of a group $\boldsymbol{G_S}$ exists then $\boldsymbol{G_S}$ is a sustainable one by a goal Q, or by a predicate Q.

The same property can be formulated in a more general form.

The goal of a system S can be defined with the help of an unary or a n—ary predicate Q in such a way:

$$\bigwedge_{i \in I} P_i \Rightarrow Q$$

or

$$\bigwedge_{i \in I} P_i(S) \Rightarrow Q(A_1, \ldots, A_n)$$

where $A_1, \ldots, A_n$ are in the domain of a predicate Q. The numerical characteristics or the structural characteristics of the system can change but at the same time the system S' which was obtained as a result of these changes can satisfies the predicate P that defines the integrity property of a system S. In this case we have:

Fig. 3.7 Cyclic group of order 3

$\circ$	e	a_1	a_2
e	e	a_1	a_2
a_1	a_1	a_2	e
a_2	a_2	e	a_1

$P(S) \Rightarrow P(S')$, and so a system S' fulfils al the functions as a system S, and as a consequences of it reaches the goal Q. Thus, quantitative changes or structural changes within certain limits do not violate the integrity property of a system. This property will be called quasi-sustainability on a predicate Q of a system S.

In the Chaps. 6 and 10 we shall show while studying P-innovative, P-pseudo-innovative and P-effective systems that all these properties are in deep connection with the notion of a sustainability of a system. Now consider Fig. 3.7.

Let's consider the basic relations of this group:

$$a_1 \circ a_1 = a_2 \tag{3.2}$$

$$a_2 \circ a_2 = a_1 \tag{3.3}$$

In the case (1) the following options are possible:

(1) indicators which characterize a_1 and a_2 are of the same nature of monotony,
(2) indicators which characterize a_1 and a_2 are of the different nature of monotony.

In the case (2) the following options are possible:

(3) indicators which characterize a_1 and a_2 are of the same nature of monotony,
(4) indicators which characterize a_1 and a_2 are of the different nature of monotony.

In (3.2) and (3.3) if (2) or (4) takes place then the model $\boldsymbol{G_S} \cong \boldsymbol{Z_3}$ outlines the possible crisis. If so then the algorithm is finished. Otherwise go to step 2.

2 step. If in (3.2) and (3.3) we have respectively (1) and (3) then the model $\boldsymbol{G_S} \cong \boldsymbol{Z_3}$ does not outline the possible crisis.

In accordance with diagram (3.1) the one-to-one correspondence of factors determining the system and the elements of the system takes place:

$$a_1 \leftrightarrow S_1 = \{s_i | i \in \Lambda_1\}$$
$$a_2 \leftrightarrow S_2 = \{s_i | i \in \Lambda_2\}$$

Let's consider $S \backslash (S_1 \cup S_2)$. Then choose $a_3 \leftrightarrow \{s_i | i \in \Lambda_3\}$, where $S \backslash (S_1 \cup S_2) \cap \{s_i | i \in \Lambda_3\} \neq \emptyset$, and consider the minimal by inclusion group $\boldsymbol{G}_S^{(2)}$, which contains a_3. After that in Cayley table T_S^2 of a group $\boldsymbol{G}_S^{(2)}$ we look for relation detecting possible crisis, that is equalities of the form $a_i \circ a_j = a_k$ such that a_i and a_j

have numerical indicators of the same nature of monotony not coincides with the nature of monotony of numerical indicator of a_k. In the case when such equality exists the model $\boldsymbol{G}_S^{(2)}$.

3.7 Loss Detection Point of Sustainability of a System Algorithm that Uses Models of Groups of Factors Describing the System

When modeling any system, the question of how to determine possible points of crisis in their functioning arises. The main question which arises here is the following one. Let $\boldsymbol{G}_S$ be a model of algebraic formalization of a system S not detecting or in other words not noticing the onset of the crisis point in the development of a system. Let $\boldsymbol{G}_S$ be a model of algebraic formalization of a system S, which not detects or in other words not notices the onset of the crisis point in the development of a system. The question of how we should change or supplement the model $\boldsymbol{G}_S$ in order that it would be able to predict or to "see" the onset of a possible crisis arises. The theory of catastrophes and the theory of bifurcations gives an answer to this question for continuous models.

We propose to use the following algorithm in our case, the case of discrete models of algebraic formalization of smart systems.

Let again a system S consists of following elements: $S = \{s_\propto | \propto \in \Lambda\}$. Let the main set of a group of factors determining the system S, $G_S = \{e, a_1, \ldots, a_n\}$, and the elements of a system S which define or one-to-one correspond to each factor are highlighted: $a_i \leftrightarrow S_i = \{s_{\propto_i} | \propto_i \in \Lambda_i,\ i = 1, \ldots, n\} \neq \emptyset$. Let us note that each subset $S_i = \{s_{\propto_i} | \propto_i \in \Lambda_i,\ i = 1, \ldots, n\}$ of a set S one-to-one corresponds the subset $\{\Lambda_i | i = 1, \ldots, n\}$ of a set Λ.

Let $F_S = \{f_i | i \in I\}$ be functions which corresponds to the set of factors $\{a_i | i \in I\}$ determining the system S. As it was mentioned above we can assume without loss of generality that $\{a_i | i \in I\}$ coincides with $G_S = \{e, a_1, \ldots, a_n\}$.

1 step. Chose any three factors e, a_1, a_2 determining the closed subsystem of a system S and consider a model $\boldsymbol{G}_S \cong \boldsymbol{Z}_3$ with the main set $G_S = \{e, a_1, a_2\}$: outlines the possible crisis, and in this case the algorithm is finished. If not then we make the next similar iteration.

References

1. Kurosh, A.G.: Theory of Groups. Nauka, Moscow (1967). (in Russian)
2. Serdyukov, V.I., Serdyukova, N.A., Shishkina, S.I.: Modeling of compensatory sustainability on the basis of algebraic formalization of systems. In: International Scientific Conference Fundamental and Applied Problems of Mehanics (FAPM—2017), LNCS, pp. 178–179.

Publishing house MSTU. N.E. Bauman, Moscow (2017). http://fn.bmstu.ru/en/international-conference-fs-en/item/616-fundamental-a. (in Russian)
3. Serdyukova, N.A., Serdyukov, V.I., Uskov, A.V., Slepov, V.A., Heinemann, C.: Algebraic formalization of sustainability in smart university ranking system. In: Jain, L.C., Howlett, R. J., Uskov, V.L. (eds.) Smart Innovation, Systems and Technologies Book Series (SIST, vol. 75) International Conference on Smart Education and Smart E-Learning, pp. 459–474 (2017)
4. Demidovich, B.P.: Lectures on Mathematical Theory of Sustainability. Nauka, Moscow (1967). (in Russian)
5. Nogin, V.D.: Theory of stability of motion. Faculty of PMPU, St. Petersburg, St. Petersburg State University (2008). (in Russian)
6. Bratus, A.S., Novozhilov, A.S., Rodina, E.V.: Discrete dynamical systems and models in ecology. Moscow State University of Railway Engineering, Moscow (2005). (in Russian)
7. Smale, S.: Book reviews. Bull. Am. Math. Soc. **84**(6), 1360–1368 (1978)
8. Arnold, V.I.: Theory of Catastrophes, 3rd edn., Science, Moscow (1990)
9. Arnold, V.: Problems in present day mathematics. In: Browder, F.E. (ed.) Mathematical Developments Arising From Gilbert Problems, Proceedings of Symposia in Pure Mathematics vol. 28, American Mathematical Society, p. 46 (1976)
10. Prigozhyn, I., Stengers, I.: Order From Chaos. A New Dialogue Between Man and Nature. Progress, Moscow (1986). (in Russian)

Chapter 4
External and Internal Properties of a System. Integrity and *P*-Integrity of a System by Predicate *P*. Formalization Smart Systems' Axiomatic

Abstract We begin this chapter from the consideration the basic properties that determine the system: integrity, internal and external attributive features that determine the system, that is, distinguishing precisely this system from all others, in the language of algebraic formalization. Then the integrity property is generalized to the case of *P*-integrity and *P*-internal and *P*-external attributive characteristics of the system, allowing to classify the properties of the system according to their various components. The notions of a quasi-stable system with respect to the property *P* and the innovation system with respect to the property *P* are introduced. While constructing the algebraic formalization of smart systems, we shall take into account the Gödel incompleteness theorem, the essence of which is that it is impossible to describe the system by using the means of this system only. As well a hierarchy of different levels links of the system is constructed. A theorem on the description of the system's links is proved. In addition, on the base of the theory of binary relations constructed by A. I. Mal'tsev, a classification of the binary relations of a system of each finite level is upbuild. The obtained results are applied to construct smart system's axiomatic and to models describing the system's synergistic effects and the processes of system's decomposition and synthesis.

Keywords *P*-integrity · Quasi-stable system · Synergetic

4.1 Introduction

The theory of systems is interdisciplinary, and it uses a high level of abstraction. The abstract theory of systems numbers nine levels: symbolic or linguistic, set-theoretic, abstract-logical, topological, logical-mathematical, information-theoretic, dynamic, and heuristic and synergistic [1]. Of course, this division is very conditional and not clear enough. In this chapter, we consider only the basic properties that determine the system: integrity, internal and external attributive features that

N. Serdyukova and V. Serdyukov, *Algebraic Formalization of Smart Systems*,
Smart Innovation, Systems and Technologies 91,
https://doi.org/10.1007/978-3-319-77051-2_4

determine the system, that is, distinguishing precisely this system from all others, in the language of algebraic formalization.

The system approach makes it possible to set and solve two problems:

- expand and deep the notion of the interaction of objects in the system; to study and, perhaps, to discover its new properties by examining the structure of its internal and external links;
- to change the efficiency of the system in various directions of its functioning, which makes it possible to manage the system.

In this chapter the integrity property is generalized to the case of P-integrity and P-internal and P-external attributive characteristics of the system, allowing to classify the properties of the system according to their various components. The formalization of the system goal made it possible to introduce the notion of a quasi-stable system with respect to the property P and the innovation system with respect to the property P. The essence of the system approach runs as follows: all the elements of the system and all operations in it should be considered only as one whole, only as an aggregate, only in interrelation with each other. Moreover, in constructing the algebraic formalization of smart systems, we shall fully take into account the Gödel incompleteness theorem, the essence of which is that it is impossible to describe the system by using the means of this system only. Therefore, to formalize smart systems, we apply a factor approach, see Definition 4.2. Besides it, in this chapter we introduce the notion of external attributive features of the system and internal attributive features of the system, with the help of which we shall formalize the axiomatic of smart systems. As well a hierarchy of different levels links of the system is constructed. A theorem on the description of the system's links is proved. Then, on the base of Mal'tsev's theory of binary relations, a classification of the binary relations of a system of each finite level is upbuild. After that we apply the obtained results to models describing the system's synergistic effects and the processes of system's decomposition and synthesis.

4.2 System Approach Basic Principles. System's Links. Connection with Synergetics

The concept of equilibrium plays the important role when studying almost every discipline, since it is this concept that determines the starting point of the change in the behavior of the system. Another concept of studying the system approach is the synergetic concept and it is directly opposite to the equilibrium concept. The emergence of a synergetic concept is associated from one side with the creation in the 1970s the general physical theory of critical phenomena and from the other side with a review of the possibilities of a dynamic approach to the description of physical systems. Such a revision was necessary, since it became clear that the application of dynamic research methods to the systems of many interacting subsystems is impossible. The revision was based on the rejection from a fully deterministic[1] description of the systems of many interacting subsystems and on the transition to a partially deterministic description using a small number of parameters. In addition, the concept of probability prediction of system's behavior was used to describe the functioning of the system, since the laws of random processes and phenomena and the laws of uncertainty are used in synergetics, and it is believed that order and chaos are interrelated and generate each other. However, the dynamic approach has an impact in this case also.

It is discovered in the fact that the probabilistic nature of the behavior of the system is explained on the basis of the incompleteness of its description, that is, the assumption of the existence of hidden dynamic parameters that obey a more accurate dynamic theory. In Chap. 2, we have already considered dynamic predicates and a probabilistic change in the model of the system's description using the groups of factors that determine the system in this connection. It is established empirically that the behavior of a real dynamic system is more like a chaotic, random behavior. Indicator of chaos or unorganized environment is homogeneity, amorphousness, sustainability. Information assessment of the chaos or unorganized environment corresponds to the maximum of entropy.[2] The maximum of entropy reveals itself in the weak sensitivity of the environment to the action of various kinds of perturbations: energy, information. At the same time, the production[3] external influences quickly fade out, without giving a noticeable effect. The properties of the unorganized medium at the present time make it possible to characterize its state only by stable deterministic probabilistic estimates, with the help of which it is impossible to predict the development of the system. In connection with the randomness and chaotic changes, questions about the feasibility and about mechanisms of reliable long-term forecasts in various areas of the life of systems arise.

Let us recall an example from the field of economics. The question about the possibility of building long-term reliable forecasts in the economy has always

---

[1]Determinism is a theory according to which the results of the development of the system are completely predetermined. A system of deterministic chaos is a system that demonstrates seemingly random results, even when these results are generated by a system of equations.

[2]In control theory, entropy is a measure of the uncertainty of the state or behavior of a system under given conditions.

[3]In the most general form, the production system is specified using a set of rules of the following type:

If S, then R, otherwise...

...

If S, then R,

where:

S is a description of some situation,

R is the set of actions that must be performed in this situation.

The production system provides process control based on the comparison with the sample. Structurally, the system consists of a set of production rules, working memory and a "recognition-action" control cycle.

occupied a special place. In the works of the 2004 Nobel Prize winners in economics, Kidland and Prescott, it is shown that if long-term goals prevail in state economic policy, then they yield a better result than when short-term goals which seem successful and the "immediate response" policy prevail.

The task of implementing long-term plans is much more complicated than the implementation of short-term and medium-term programs. To change the state of chaos a different path may be chosen in addition to the force effect on the entire environment. This path is a violation of the state of equilibrium in local areas of the environment, with the help of the introduction of order and organization. Such an effect will lead to a decrease in entropy and the processes in the ordered regions will become more correlated with the initial conditions. This will create the prerequisites for predicting events in ordered areas. Chaos is structuring and zones with a stable state of constituent units appear. The methodology of dynamics is focused on the study of motion as a consequence of the causes, which are generating or changing the motion. In the Theory of Systems, motion is understood as a change in state due to external and internal causes. Thus, the movement is a manifestation of the internal processes of the system and the influence on the system of external factors. The movement of the system is its most significant characteristic, since it fully reveals the properties of the system and allows to correlate its state with the desired goal. This leads to the following task: to influence the system, or to control the system so as to bring it to the required state. Therefore, the dynamic properties of the system are characterized by changing the parameters or factors which determine the state of the system. Dynamic instability can play and a constructive role in the functioning of open systems. Chaotic behavior is manifested as by conservative systems, i.e. systems in which the energy is conserved, and as by dissipative systems, that is, in the open systems which function in a stable state that arises in a nonequilibrium environment under the condition of dissipation (dissipation) of energy coming from outside. The dissipative system is characterized by the spontaneous appearance of a complex, sometimes chaotic structure. The essence of systems' change is based on the occurrence of irreversible processes due to instability.

After the formation of stable structural elements from chaos, their development begins:

- growth;
- the appearance of new functional qualities.

The accumulation of new properties is associated with bifurcations or the appearance of a qualitatively different behavior of the element during the quantitative change of its parameters [2]. It is assumed that at the time of bifurcation that is a kind of regeneration the probability of reliable prediction of new properties is small. In the process of systems' development contradictions arise in a natural way and they are the reason for the betterment of systems.

From the theory of systems it is known that it is impossible to speed up the development of the system by artificially introducing contradictions into it, since it is impossible to determine whether the system, as a result of their resolution, will bear the new qualities. The foregoing explains the nature of the presentation of Chap. 4. The formalization of the axiomatic of smart systems we begin with a review of the basic principles of the systems approach, the study of which requires the use of a synergistic approach. Let us now outline the main principles of the system approach, the study of which is hardly possible without taking into account to some extent the synergetic approach.

These are the following principles:

- the aggregate of the system's elements is considered as one whole, possessing a set of definite links and properties. So it turns out that the system is not a simple union of its elements. It is necessary to take into account the links between the elements of the system, providing certain properties of the system, that is, the structure of the system;
- the properties of the system are not simply the sum or the union of the properties of its elements. The system can have special properties, which may not exist for the individual elements that arise due to the connections between the elements of the system, that is, due to the structural links of the system. The researcher, using a system approach, first decomposes the system into subsystems and elements, determines the goals of their functioning, the criteria for evaluating their effectiveness, builds models for their functioning, and then sequentially synthesizes them into the system model. This problem is extremely complicated, because the structure of system's links is not known a priori to the researcher. The structure of the system's links is closely related to synergistic effects. In [3] the history of the origin of the synergetic approach is given. Possible sources of synergetics are the tectology of A. A. Bogdanov, the general theory of systems of L. von Bertalanffy, and cybernetics, founded by N. Wiener. The essence of synergetics as a science was defined by the professor of the Institute of Synergetics and Theoretical Physics in Stuttgart G. Haken in 1973: "Synergetics is the direction of interdisciplinary research whose objects are the processes of self-organization in open systems of various nature. In such systems, far from thermodynamic equilibrium, due to the flow of energy and matter from the external environment, nonequilibrium is created and maintained."

Haken identified key aspects that express the main content of synergetics [4]:

(1) systems consist of several or many identical or dissimilar parts that are in interrelation with each other;
(2) systems are nonlinear;
(3) the systems under consideration are open, we examine systems which are far from thermal equilibrium. We note that in the context of this book we are talking about closed systems of factors which determine the system S being studied, that is, we just assume that the interaction of any finite number of factors which determine the system does not go beyond this system. We note that in the context of this book we are talking about closed systems of factors which determine the system S being studied, that is, we just assume that the

interaction of any finite number of factors which determine the system does not go beyond this system. In System Theory an open system is understood as a system that continuously interacts with its environment. At the same time, the interaction can take the form of information, energy or material transformations at the border with the system. An open system is opposed to an isolated system that does not exchange energy, substance or information with the environment. We are talking about the factors that determine the system.

(4) systems are subject to internal and external fluctuations;
(5) systems can become unsustainable;
(6) qualitative changes occur in systems;
(7) emergent new qualities appear in systems;
(8) arise spatial, temporal, spatio-temporal or functional structures;
(9) structures can be ordered or chaotic;
(10) mathematization is possible in many cases.

Paragraph (8) we reformulate as follows: there are spatial, temporal, spatio-temporal or functional structures of the system's links. Paragraph (9) we reformulate as follows: the structure of the system's links can be ordered or chaotic, that is, not to have connections. Paragraph (10). The present book is devoted to the mathematization of Smart Systems Theory. In 1937, the Nobel laureate, Academician L. D. Landau published the papers "Theory of phase transitions" and "On the theory of phase transitions", which explained the abrupt change in the properties of matter under certain conditions [5]. A phase transition is a transition of matter from one thermodynamic phase to another, for example, boiling of water or condensation of steam. In these works, Landau first introduced the term used in modern synergetics, an order parameter by which one can describe a new state of matter as a result of a transition from an unordered phase to an ordered one. The value of the order parameter equals zero on one side of the phase transition, and is a finite number on the other side of the phase transition. In this case, with a small change in the order parameter, a disproportionate change in the properties of the substance takes place. From the point of view of modern synergetics, the theory of phase transitions extends to any system. In this case, the order parameter is an indicator of the synergistic effect. In 1948, Professor of the Massachusetts Institute of Technology of the United States N. Wiener published a book "Cybernetics or management and communication in living organisms and machines" [6]. In this book, he examined the behavior of technical and physiological systems from the standpoint of behaviorism which is the science of behavior. The behavioral method consists in examining the output of the object and the relationship between the output and the input. An output is any change made by an object in an environment. An input is understood as any event external to the object, changing in any way this object. Wiener noted the similarity of control and communication processes in living organisms and machines and determined the important property of such systems that is purposefulness. The integrity property (or purposefulness, goal fullness) in fact, is the determining factor for the system. It is this property that distinguishes the system from a simple aggregate of its constituent elements. At the same time, Wiener focused on the processes of transmission, storage and processing of information [6]. All of the above, the first and second principles, talk about the importance of links in the system, and therefore we shall first concentrate on the hierarchy of the system's links.

4.3 The Model of Hierarchy of Structural Links of the System

In this section we describe the hierarchy of structural links of the system.

Let's designate:

S—the designation of the system;

$S_0(S)$—elements of the system;

$C_0(S)$—links between elements of the system, that is between the elements of the set $S_0(S)$.

Since the links are the relationships between the elements of the system, for binary links of the first level of the system S, i.e. for connections between two elements, we have:

For a binary link $\propto$ between the elements of the system: $\propto \in S_0(S) \times S_0(S)$.

For n-ary link of the first level of the system S we have:

$\propto \in S_0(S) \times S_0(S) \times \cdots \times S_0(S)$, where the number of factors on the right-hand side is n.

$P_1(S) = \{S_0(S); C_0(S)\}$ is a two-element set whose elements are the set of elements of the system S and the set of links $C_0(S)$ of the elements $S_0(S)$ of the system S.

Let $\boldsymbol{P}_1(\boldsymbol{S}) = \langle P_1(S), \propto_1 \rangle$, where $\propto_1$ is a partial order on the set $C_0(S)$, that is $\propto_1$ structures the set of links $C_0(S)$ of the elements of the set $S_0(S)$ (we can assume so, because one can go from a partially ordered set to the lattice in accordance with [7]).

We call $\boldsymbol{P}_1(\boldsymbol{S}) = \langle P_1(S), \propto_1 \rangle$ the structure of the constraints of the first level of the system S.

Let us construct the second-level links of the system.

$P_2(S) = \{P_1(S); C_1(S)\}$ is a two-element set whose elements are sets $P_1(S)$ and $C_1(S)$, $C_2(S)$ are the links between elements of the set $C_1(S)$, that is $C_2(S)$ structures links of the level 1 of the system S.

$C_2(S)$ are the links of the level 2 of the system S.

For the second-level links of the system S we have:

- for binary relation $\propto_2$ we have: $\propto_2 \in C_1(S) \times C_1(S)$,
- for n-ary relation $\propto_2$ of the second level of the system S we have:

$\propto_2 \in C_1(S) \times C_1(S) \times \cdots \times C_1(S)$, where the number of factors on the right-hand side is n.

Let us suppose that at the step n links' structure $\boldsymbol{P_n(S)} = \langle P_n(S), \propto_n \rangle$ of the system S of the level n has already been constructed. Here $P_n(S) = \{P_{n-1}(S); C_{n-1}(S)\}$ is a two-elements set, whose elements are the sets $P_{n-1}(S)$ and $C_{n-1}(S)$, and $\propto_n$ is a binary relation of a partial order on the set $C_{n-1}(S)$, that is $\propto_n$ structures the set of links $C_{n-1}(S)$ of the set $P_{n-1}(S)$, $C_{n-1}(S)$ are the links of the level $n-1$ of the system S. Elements of the set $C_n(S)$ are the links of the elements of the set $C_{n-1}(S)$; $C_n(S)$ are the links of the level n of the system S.

Then $P_{n+1}(S) = \{P_n(S); C_n(S)\}$ is a two-element set whose elements are sets $P_n(S)$ and $C_n(S)$, $C_{n+1}(S)$ are links of the elements of the set $C_n(S)$, and $\propto_{n+1}$ is a relation of a partial order on the set $C_{n+1}(S)$, that is $\propto_{n+1}$ structures the links of the level n of the system S. $C_{n+1}(S)$ are the links of the level $n+1$ of the system S.

So we obtain that the links of the level k of the system S are defined for any natural number by the axiom of mathematical induction.

Definition 4.1 The system $\boldsymbol{P_1(S)} = \langle P_1(S), \propto_1 \rangle$ will be called the first level system dual to the system S.

Let us illustrate this with the help of the following scheme.

Let S be any system, $S_1(S) = \{a, b, \ldots, c, \ldots\}$ be the set of elements of the system S, $C_1(S) = \{v, w, \ldots, z, \ldots\}$ be the set of links between elements of the system S. Links of the first level of the system can be illustrated with the help of the graph of the system:

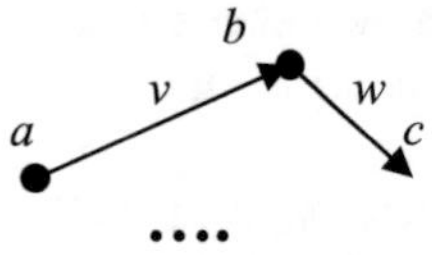

b
v w
a c (4.1)
....

The edges of the graph (4.1) are the elements of the set $C_1(S)$, representing the links of the elements of $S_1(S)$ of the system S. At the second level we consider the links $C_2(S)$ between the elements of the set $C_1(S)$, which can also be represented by a graph:

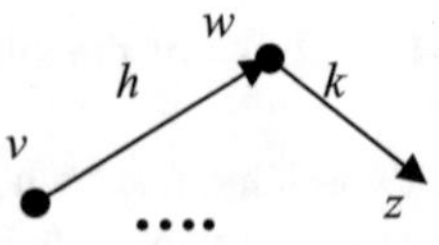

w
h k
v (4.2)
.... z

This process is then continued by induction.

Let's give the following example from the field of smart systems, which justifies the consideration of links of different levels in systems and having been used in e-learning. Really interdisciplinary links reveal the links between the relationships of properties in different subject areas. Besides it, the connections between the properties of the system justify and explain the meaning of the introduction of the P-properties of the system and, ultimately, the introduction of the purities by the predicates [8].

A partial description of the connections of a system of levels not exceeding n enables us to obtain the following theorem.

Theorem 4.2 (about the description of the system's links)

Links of the level no more than n of the system S, where n is a natural number, are determined by no more than two combinations of connections of the level no more than n of the system S.

Proof Let's designate by $\{a_1, a_2, \ldots, a_n\}$ the set of links of the level no more than n of the system S. Let's consider the free group of factors $F_n = \langle a_1, a_2, \ldots, a_n \rangle$ of the rank n with the generators $a_1, a_2, \ldots, a_n$. Every free group of a finite rank can be embedded into a free group of countable rank, the commutant $[F_2, F_2]$ of a free group F_2 of the rank 2 is a free group of a countable rank, so we can embed $F_n = \langle a_1, a_2, \ldots, a_n \rangle$ into a commutant $F_\infty = [F_n, F_n]$, which is a free group of countable rank, after that a commutant $F_\infty = [F_n, F_n]$ should be embedded into a commutant $[F_2, F_2]$ of a free group F_2 of a rank 2. So we have that there exist two words $w_1(a_1, a_2, \ldots, a_n)$ and $w_2(a_1, a_2, \ldots, a_n)$ such that $F_n = \langle a_1, a_2, \ldots, a_n \rangle$ can be embedded into the group $F_2 = \langle w_1(a_1, a_2, \ldots, a_n), w_2(a_1, a_2, \ldots, a_n) \rangle$.

The approach of allocating connections of various levels in the system is justified, since:

- the system approach is applied in the case when it is a question of the description of complex systems, i.e. when the behavior of the system cannot be described with the help of a single mathematical model,
- there are numerous examples showing the existence of such links in different systems, and in particular in Smart Systems.

4.4 Types of System Connections. Different Types of Classifications. Classification of Binary Links of the First Level of the System

Definition 4.3 Let $\propto \in S_0(S) \times S_0(S)$ be a binary link of the first level of the system S. A link $\propto$ is called a reflexive one if the binary relation $\propto$ is a reflexive one on the set $S_0(S)$, that is

$$(\forall a \in S_0(S))(a \propto a \, is \, true).$$

A link $\propto$ is called a symmetric one if the binary relation $\propto$ is a binary symmetric relation on the set $S_0(S)$, that is $(\forall a, b \in S_0(S))(a \propto b \Rightarrow b \propto a)$.

A link $\propto$ is called an antisymmetric one if the binary relation $\propto$ is an antisymmetric binary relation on the set $S_0(S)$, that is

$$(\forall a, b \in S_0(S))(a \propto b \,\text{and}\, b \propto a \Rightarrow a = b).$$

A link $\propto$ is called a transitive one if the binary relation $\propto$ is a transitive binary relation on the set $S_0(S)$, that is $(\forall a, b, c \in S_0(S))(a \propto b \,\&\, b \propto c \Rightarrow a \propto c)$.

A binary link $\propto$ is called an equivalence if the binary relation $\propto$ is a binary relation of equivalence on the set $S_0(S)$ that is $\propto$ is a reflexive, symmetric and transitive binary relation on the set $S_0(S)$.

It is possible to eliminate the link α of the first level of the initial system S by considering the factor-system $S_0(S)/\propto$ with respect to the equivalence relation $\propto$ of the first level of the system S. The same operation can be done for any finite level of the system.

4.4.1 Operations Over System Links

Since we consider the connections of the system S as a relation between the elements of some set $C_n(S)$, this approach allows us to use the main results of AI. Malt'sev [7], which concern operations over relations. Hence we obtain:

1. The set of all binary links of a fixed level (realizable and unrealizable by the system under consideration) forms a Boolean algebra with respect to operations of union, intersection and taking complement of binary links. One can also speak of operations of inversion and multiplication of links in addition to operations $\cup, \cap, '$ over binary links of a fixed level.
2. Let us recall the definition of the products of relations and the inversion of relations [7]. Let the relation $\propto$ be defined on a pair of sets A, B and the relation β is defined on a pair of sets B, C. The product $\propto \cdot \beta$ of relations $\propto, \beta$ is called a relation defined on a pair of sets A, C such that $a \propto \cdot \beta c$ is true if and only if there exists an element x in the set B, such that $a \propto x$ and $x \beta c$ are true. If the relation $\propto$ is given on a pair of sets A, B, then the inverse relation or the inversion of a relation $\propto$ is called a relation $\propto^{-1}$, which is defined on a pair of sets B, A, consisting of all pairs (b, a), for which $(a, b) \in \alpha$.

Hence from we obtain the following assertion.

Theorem 4.4 *Let S be a system and $C_n(S)$ be a set of links of the level n of the system S. An algebra $\mathbf{C}_n(S) = \langle C_n(S), \cdot, \square^{-1}, e \rangle$, where $e = \{(a, a) | a \in C_n(S)\}$ is a group if $(\forall a, c \in C_n(S))(\exists x \in C_n(S))(a \propto x \,\&\, x \beta c)$ for every $\propto, \beta \in C_n(S)$.*

This group is called a group of all links of the level n of a system S.

Different predicates P, given on the class of groups and closed under taking subgroups and factor-groups, determine the properties of the links of any fixed level of the system S, if these links themselves form a group. Since the number of synergetic effects of a closed associative system with feedback is determined by its number of final states, then if the links of a system of some finite level k themselves form an associative closed system with a feedback of n_S^k elements, then the number of synergies of the system S is not exceeds the number of synergetic effects of the closed associative structure of the level k of the system S with a feedback if the number of pairwise nonisomorphic groups of order n_S^k is greater than the number of pairwise nonisomorphic groups of the order n_S. Here from one can obtain a classification of the links of a closed associative system with feedback in terms of the number of synergetic effects of the system S and the number of synergetic effects of the closed associative structure of links of the level k of the system S with a feedback.

4.5 Closed Associative Systems with Feedback. Partial Classification on the System Links Levels and the Number of Synergistic Effects

Undoubtedly, the construction of an exhaustive detailed classification of closed associative systems with a feedback even over the levels of the system's links is hardly possible at the present time. Therefore, we consider a special case. Let us suppose that a group of factors G_S, which determines the system S, is finite. We also assume that for every positive integer n the set of links of the level n has a group structure, that is $\boldsymbol{C_n(S)} = \langle C_n(S), \cdot, \square^{-1}, e \rangle$ is a group. Let us recall once again that in this book we are talking about a closed systems of interacting factors that determine the system, and not a closed or open systems in the classical sense of Theory of Systems.

Let's designate:

$g(n)$ is a number of pairwise non isomorphic subgroups of a group of the order n. For example, if p is a prime number, then $g(p) = 1$.
A partial classification will be made on specific examples which show how one can act in the general case within the framework of the assumptions made.

Let's consider the following example.

Let the group of factors G_S, which determine the system S, consists of p elements, where p is a prime number, $p > 2$. Then G_S defines the model of the system S without synergistic effects. Let's suppose, that the system S is arranged in such a way that each of its elements is connected with each other, that is that $|C_1(S)|$ the number of the links of the first level of the system S is not less than $p \times (p - 1)$. Besides it, let's assume that the links of each level k of the system S, where k is a natural number, also satisfy this condition, that is, the links form a group, and each element of $C_k(S)$ is connected with each other, that is, for each natural number n the following relation holds:

$$x_{n+1} = x_n(x_n - 1)$$

where $x_n = |C_n(S)|$. Then the number of possible synergistic effects of the links of the system of each level k, where k is a natural number, is greater than 1. If $p = 2$, then the model G_S is the simplest model without synergistic effects at all levels of links. If, at some level of the links of the system S in the G_S model, we arrive at the situation where $|C_k(G_S)| = p = 2$, then the subsequent link levels of the system S do not contain synergistic effects according to the G_S model. These examples show that $p = 2$ plays a special role in the theory of systems, as well as in algebra (field theory), as in the theory of numbers. Introduction of the concept of factors that determine the system, allows us to introduce the notion of factor fractality in terms of links levels of the system.

Let's make the following remark. It is possible to classify the finite models $\boldsymbol{G_S}$ of factors which determine the system S, in the case when for each positive integer n the links of the system of level n has a finite group structure, that is, $\boldsymbol{C_n(S)} = \langle C_n(S), \cdot, \square^{-1}, e \rangle$ is a group, since a complete description of finite groups has now been obtained, see Atlas of Finite Groups by J.H. Conway, R.T. Curtis, S.P. Norton, R.A. Parker, R A. Wilson, Clarendon Press, Oxford, 1985.

Definition 4.5 The system S is called factor-fractal by levels i, j, if the group of links $G_i(S)$ of level i is isomorphic to the group of links $G_j(S)$ of level j of this system.

Such a fractality we encounter, for example, in biology when transferring properties from parents to offspring.

4.6 System Binary Links and Mappings

With the help of [7] one can single out the functional links of any fixed level of the system S.

Definition 4.6 A binary link of a level n of a system S is called a functional binary link of a level n of the system S if

$$(\forall a \in C_n(S))(\exists! b \in C_n(S))(a\alpha b)$$

Note that when constructing a mathematical model of a system structural links, optimization of the choice of dominant links is necessary, since an adequate analysis of a huge number of parameters is hardly possible. In part, this explains our approach to describing the system with the help of factors that determine the system.

The next principle of the Theory of Systems and System Analysis prohibits to consider the system in isolation from its environment. This means that the analyzed system is considered as a structural part that is a subsystem of some more general system.

Decomposition

The principle of taking into account the external environment of the system leads to the principle of dividing the given system into subsystems, that is, to the principle of decomposition. The principle of taking into account the external environment of the system leads to the principle of dividing the given system into subsystems, that is, to the principle of decomposition. If the subsystems which are received as a result of decomposition process are not available for analysis, then the decomposition process is applied to them again. At the same time, in the process of decomposition one can not violate the remaining principles of the system approach, if we want to ensure the consistent applicability of algorithms to solve System Analysis problems. **Decomposition is the division of the whole into parts.** In addition, decomposition is a scientific method which uses the structure of a problem to replace a solution of one big problem by solving a series of smaller, interrelated tasks, simpler than the original ones.

Decomposition, as a process of separation, allows us to treat any investigated system as a complex one, consisting of separate interconnected subsystems, which, in turn, can also be disconnected. Herewith not only material objects, but also processes, phenomena and concepts can act as systems. An important role in the process of decomposition is played by the split-off the links levels of the system, since the separation of links levels in the system is in fact decomposition.

Let's now build an algorithm for the system decomposition by its links levels.

4.7 Algorithm of Analysis and Decomposition of the System by Its Links Levels

Step 1. Single out the elements $S_0(S)$ of the system S.
Step 2. Single out the links $C_0(S)$ between the system elements that is between the elements of the set $S_0(S)$.
Step 3. Investigate the structure of the first level links $\boldsymbol{P}_1(\boldsymbol{S}) = \langle P_1(S), \propto_1 \rangle$, where $\propto_1$ is a partial order relation on the set $C_0(S)$.
Step 4. Continue this process by induction on the levels of the system links in accordance with the model of the hierarchy of structural links of the system.
Step $3n$. Step $3n+1$. Step $3n+2$. Construct the structure of the links $\boldsymbol{P}_n(\boldsymbol{S}) = \langle P_n(S), \propto_n \rangle$ of the level n of the system S, where $\propto_n$ is the partial order relation on the set $C_{n-1}(S)$.

By the axiom of mathematical induction, this process can be continued to any set of links of the level n where n is a natural number.

At the same time, the direction of the algorithm for decomposing the system from the bottom to the top is from the lower level of communication to the upper one.

4.8 Example. System Decomposition. Smart System The World University Rankings. Evaluation of the The World University Rankings System

This system uses 13 parameters (or, evaluation criteria) with weights, which are expressed in percentages from the total score on the following several categories of evaluation criteria:

Category I: "Teaching and learning environment"—30% upon parameters:

1. The survey of the scientific university staff's environments—15%;
2. The ratio of the number of professors and teaching staff to the number of students in high school—4,5%;
3. The ratio of the number of masters who completed a degree Ph.D., to the number of bachelors enrolled on masters—2,25%;
4. The ratio of the number of professors and teaching staff with Ph.D. degree to the total number of professors and teaching staff—6%;
5. The ratio of an income of an educational institution from the scientific research to the number of scientific staff—2,25%;

Category II: "Research—volume, income, reputation"—30% upon parameters:

6. Research reputation among similar universities—18%;
7. Income from research activities, correlated with the number of employees and normalized for purchasing power parity—6%;
8. Assessment of the environment for scientific research which is calculated as the ratio of the numbers of papers indexed in Thomson Reuters as scientific to the number of professors and teaching staff and then this ratio normalized;
9. Assessment of the environment for scientific development—6%;

Category III: "Citation, influence, authority"—30% upon parameters:

10. The category consists from one indicator an index of citing. The ranking does not include universities that published less than 200 papers per year;
11. Data are normalized to reflect the difference in the various scientific fields;

Category IV: "Income from production activities: innovations"—2,5% upon parameter:

12. Evaluation of knowledge transfer—innovations, consulting, invention;

Category V: "International image"—7,5% upon parameters:

13. The share of foreign undergraduate and graduate students in relation to local students and postgraduates—2,5%;
14. The share of foreign citizens among the professors and teaching staff in relation to the total number of professors and teaching staff—2,5%;
15. The share of scientific publications, which have at least one foreign co-author relative to the total number of publications of the University—2,5%.

There are five categories (or, blocks) in this ranking system; as a result, in accordance with the consequence 2.3 above, this ranking is a rather sustainable one.

Let's consider a system **S** which represents The World University Rankings. A decomposition of this system gives 5 subsystems, namely S_1, S_2, S_3, S_4, S_5; they correspond to each of the five mentioned above categories. Let G_S be a group of factors, which represent the system S. Let B_1, B_2, B_3, B_4, B_5 be respectively groups of factors, which represent subsystems S_1, S_2, S_3, S_4, S_5. We may apply the additional restriction on system S and subsystems S_1, S_2, S_3, S_4, S_5—the operation of composition of the factors is a commutative one. Under this restriction a synthesis of system S is described by the following theorem from [9].

Theorem 4.7 *Let the operation of composition of factors which represent the closed associative system with a feedback be a commutative one. Then the synthesis of the systems* S_1, S_2 *is described by the group of factors* $Ext(B_2, B_1)$,[4] *the synthesis of the systems* S_1, S_2, S_3 *is described by the group of factors* $Ext(B_3, Ext(B_2, B_1))$, *the synthesis of the systems* S_1, S_2, S_3, S_4 *is described by the group of factors* $Ext(B_4, Ext(B_3, Ext(B_2, B_1)))$, *the synthesis of the systems* B_1, B_2, B_3, B_4, B_5 *is described by the group of factors* $Ext(B_5, Ext(B_4, Ext(B_3, Ext(B_2, B_1))))$. Because the numbers of factors (that represent a close associative system S with a feedback with commutative operation of composition of factors) is finite, then, in this case—to some extent, the following existing theorem allows to simplify the synthesis process of the system S.

Theorem 4.8 [10] *If* A *and* C *are finite groups then the group of extensions* $Ext(C, A)$ *is isomorphic to the group of all homomorphisms* $Hom(C, A)$, *that is* $Ext(C, A) \cong Hom(C, A)$. *For example, if* $\mathbf{G_S} \cong \mathbf{Z_m}$ *is a cyclic group of order m, then* $Hom(\mathbf{Z_m}, \mathbf{Z_m}) \cong \mathbf{Z_m}$. $Hom(\mathbf{Z_m}, \mathbf{Z_m}) \cong \mathbf{Z_m}$.

So, the synthesis process gives us exactly m different variants of the system's synthesizing. Hence, using (a) structures of analyzed university ranking systems (such as THE, QS, ARWU and GRUP systems), (b) a description of the group $Hom(C, A)$, *(c) Theorem 4.8, and (d) the final stability of blocks that make up a ranking system, we obtain estimates of sustainability of these systems, which show a rather conservative nature of these systems. It is important to note here that using Theorem 4.8, one can build a new ranking systems with a more higher sustainability in comparison with the systems discussed above.*

---

[4]*The group of extensions of an abelian group* B_1 *by the abelian group* B_2 [10].

Let us now consider the process of synthesizing the system by the levels of links of the system.

Synthesis
According to Ushakov's Explanatory dictionary [11], under the synthesis, as a method of research is understood the establishment of communication and the union of the different single elements obtained in the process of analysis into an integrated whole. Synthesis is the combination of various elements into a single whole, performed in the process of cognition and practical activity [12]. Synthesis is the process of combining or combining previously disparate things or concepts into a whole. Synthesis is the engineering construction of complex systems from pre-prepared blocks or modules of different types. Low-level, deep structural combination of components of different types. Now let's construct an algorithm for synthesizing the system by its links levels.

Algorithm 4.9 of system synthesis by its links levels After analyzing and decomposing the system using Algorithm 4.7, we climb by this algorithm in the opposite direction, that is, from top to bottom, from the last link level to the first one. This produces a tree of all possible models $\boldsymbol{G_S}$ of factors which describe the system S. Since the model $\boldsymbol{G_S}$ is finite, the number of its connections is finite and its upper, last level of connections exists.

4.9 Algebraic Formalization of the Axiomatic Description of Smart Systems

Let's begin with formalization of an axiomatic description of a system. Now we shall dwell upon the description of the integrity principle of the system and the purposefulness principle of the system.

Let us remind some useful definitions from [13–16] for the reader's convenience.

Definition 4.10 By a system we understand a two-dimensional vector

$$S = \left\langle \{\langle S_\alpha, Q_\alpha, U_\alpha\rangle | \alpha \in \mathrm{A}\}, I(S) = \left\langle \{a_\beta | \beta \in \mathrm{B}\} | \Omega_S = \left\{ f_\gamma^{n_\gamma} \middle| \gamma \in \Gamma \right\} \right\rangle \right\rangle \quad (4.3)$$

where

- $\{S_\alpha | \alpha \in \mathrm{A}\}$ is a set of all system's statuses which are possible as a result of system S operation,
- $\{Q_\alpha | \alpha \in \mathrm{A}\}$ is a set of all system's statuses Q upon which system S is affected,
- $\{U_\alpha | \alpha \in \mathrm{A}\}$ is a set of all statuses of an external environment which are possible as a result of system S operation,
- $\{a_\beta | \beta \in \mathrm{B}\}$ is a set of all internal factors that influence system S behavior.

If a composition of factors $a_1{}^\circ a_2{}^\circ \ldots {}^\circ a_{n_\gamma} = a$, than let $f_\gamma^{n_\gamma}(a_1, a_2, \ldots, a_{n_\gamma}) = a$, where $\{f_\gamma^{n_\gamma} | \gamma \in \Gamma\}$ is a set of operations on set of factors $\{a_\beta | \beta \in \mathrm{B}\}$, $f_\gamma^{n_\gamma}$–n_γ-argument operation, $I(S) = \langle \{a_\beta | \beta \in \mathrm{B}\} | \Omega_S = \{f_\gamma^{n_\gamma} | \gamma \in \Gamma\} \rangle$—an algebraic system of internal factors of system S.

Let $\{P_i | i \in I\}$ be a set of all properties of system S, which it holds as a result of its operation, $\{B_j | j \in J\}$—a set of all subsystems of system S, $\{v_m^n | m \in M, n \in N\}$—a set of all connections in system S, *—a composition operation. In this case, $G(S)$ is a goal of system S. As a result, we get an algebra $<\{P_i | i \in I\}| *>$ under the assumption that a set of all system S properties is closed under operation of composition *. In its own turn, it means that is that we have a full description of system S.

Definition 4.11 Under the algebra of factors, a system will be understood as algebra $\bar{A} = \langle A | \{f_\alpha^{n_\alpha} | \alpha \in \Gamma\} \rangle$ with a fundamental set of factors A and a set of operations $\{f_\alpha^{n_\alpha} | \alpha \in \Gamma\}$ that describe connections between factors.

Definition 4.12 A sub-algebra $\bar{B} = \langle B | \{f_\alpha^{n_\alpha} | \alpha \in \Gamma\} \rangle$ of algebra $\bar{A} = \langle A | \{f_\alpha^{n_\alpha} | \alpha \in \Gamma\} \rangle$ is called P-pure in $\bar{A}$ (or an embedding φ of a sub-algebra $\bar{B}$ into an algebra $\bar{A}$ is P-pure, if (1) every homomorphism $\bar{B} \xrightarrow{\alpha} \bar{C}$ of the subalgebra $\bar{B}$ into $\bar{C}$ (where $\bar{C}$ is an algebra of the signature $\{f_\alpha^{n_\alpha} | \alpha \in \Gamma\}$ of $\bar{A}$,) and (2) $P(\bar{C})$ is true, (3) P is a predicate on the class of algebras of the signature $\{f_\alpha^{n_\alpha} | \alpha \in \Gamma\}$ closed under taking subalgebras and factor algebras, can be continued to a homomorphism β of $\bar{A} = \langle A | \{f_\alpha^{n_\alpha} | \alpha \in \Gamma\} \rangle$ into $\bar{C} = \langle C | \{f_\alpha^{n_\alpha} | \alpha \in \Gamma\} \rangle$ in such a way that the following diagram is commutative:

$$
\begin{array}{ccc}
0 \hookrightarrow \bar{B} = \langle B | \{f_\alpha^{n_\alpha} | \alpha \in \Gamma\} \rangle & \xrightarrow{\varphi} & \bar{A} = \langle A | \{f_\alpha^{n_\alpha} | \alpha \in \Gamma\} \rangle \\
\alpha \searrow & & \swarrow \beta \\
& \bar{C} = \langle C | \{f_\alpha^{n_\alpha} | \alpha \in \Gamma\} \rangle &
\end{array}
\tag{4.4}
$$

that is $\beta\varphi = \alpha$. (A note: The general operations of the same type in algebraic systems will be denoted in identical manner.)

In fact, purities are the fractality of links. In this case, P-purities are the fractality of links with the property P.

Now we need to provide a formalization of an axiomatic description of a system.

Algebra $\bar{A} = \langle A | \{f_\alpha^{n_\alpha} | \alpha \in \Gamma\} \rangle$ with main set of factors A and a set of operations $\{f_\alpha^{n_\alpha} | \alpha \in \Gamma\}$ (that actually describe connections of factors in a system), is group $\bar{A} = \langle A |^\circ, \square^{-1}, e \rangle$, where $^\circ$ is a composition of factors (i.e. the consistent implementation of factors), $\square^{-1}$ is an operation of taking a reverse factor, and e is a neutral factor. In this case, $\bar{A} = \langle A | \{f_\alpha^{n_\alpha} | \alpha \in \Gamma\} \rangle$ is called a group of factors that represent system S.

Let us consider the meaning of purities and examples of P-purities in a class of all groups. Formula (4.4) has the following meaning in a class of groups: epimorphic images of B and A are the same in the class of all finite groups [17]. For P-purities the meaning is as follows: B and A have the same epimorphic images in class of all groups satisfying the condition P.

The possible examples include but are not limited to:

(1) P allocates the class of all finite groups in the class of Abelian groups, get the usual purity in the class of all Abelian groups;
(2) P allocates the class of all Abelian groups in the class of all groups;
(3) P allocates the class of all finite groups in the class of all groups;
(4) P highlights the diversity in the class of all groups i.e. the class of groups closed under subgroups, homomorphic images and Cartesian products, such as Burnside's variety of all groups of the exponent (indicator) n defined by the identity $x^n = 1$, the variety of nilpotent groups of class of nilpotent is not more than n, soluble groups of length not exceeding the number l, etc.

To build a formalization of the innovation system one must set the axiomatic system's description. At first let's consider the formalization of a system's goals. The system's goals can runs as follows:

1. New status S_α of a system S,
2. New status Q_α of a system Q, on which system S affects.
3. New status U_α as a result of an external agency of a system S upon outdoor environment U, $\alpha \in \mathrm{A}$.

So, we have the target vectors $\langle S_\alpha, Q_\alpha, U_\alpha \rangle, \alpha \in \mathrm{A}$, and a final status S_f of a system S, corresponding to the vector $\langle S_f, Q_f, U_f \rangle$, in which the system S moved as a result of it's functioning. The following cases are possible:

(a) $2 \Rightarrow 3$, (b) $3 \Rightarrow 2$, (c) $3 \Rightarrow 1$, (d) $1 \Rightarrow 3$.

The graph $G(S)$ the impact of targets of a system S upon the environment U runs as follows:

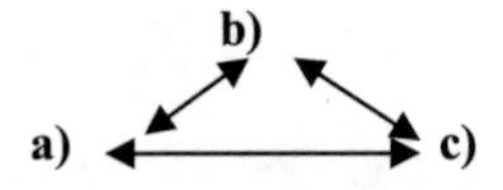

Now based on the introduced notations and concepts one can go to the formalization of an axiomatic description of the purposes and principles of the system approach. Generally adopted axiomatic (even verbal one) of the system approach or of the theory of systems does not exist.

Usually the following main principles are chose while describing the system: the presence of a target (set of targets), integrity (wholeness), hierarchy, good structure. Basic system's principles are opened in the following way. Integrity means fundamental irreducibility properties back to the sum of the properties of its constituent elements and revivalist of the last properties of the whole, the dependence of each element, properties and relationships of the system from his place, functions, etc. within the whole. Good structure means the ability to describe the system through the establishment of its structure, i.e., network connections and relations of the system, the dependence of the system behavior from the behavior of its individual elements and the properties of its structure, the interdependence of the system and environment. Hierarchy means that each component of the system can be considered as a system and the analyzed system represents one component of a broader system in this case. The adequate knowledge of the system requires building many different models, each of which describes some aspect of the system because of the principal complexity of each system. Let's describe main system's principles using narrow predicate calculus. This runs as follows.

The principle of integrity is set to items 1 and 2:

1. $\models \wedge_{i \in I} P_i(\langle S_\alpha, Q_\alpha, U_\alpha \rangle, \alpha \in \mathrm{A}) \wedge (\exists P)(P(\langle S_\alpha, Q_\alpha, U_\alpha \rangle, \alpha \in \mathrm{A}) \wedge (\neg P(\langle S'_\delta, Q'_\delta, U'_\delta \rangle, \delta \in \Delta)$ system S possesses all the properties $\{P_i | i \in I\}$, and there exists at least one property P, such that no one of own subsystems S' of the system S does not possessed P. Thus, the property of integrity allocates system as one having a synergistic effect (at least one).
2. The graph of all relations $\Gamma(S)$ of the system S is isomorphic to the graph $\Gamma(S_f)$ of all links of the system S_f—the final status of the system S.
3. The system structure is understood to be the lattice of its subsystems. Good structure principle is set to item 3:
4. Let $\{a_\beta | \beta \in \mathrm{B}\}$ be a set of all inner factors acting on the system S that is determine its behavior. If $a_1 * a_2 * \ldots * a_{n_\gamma} = a$, then $f_\gamma^{n_\gamma}(a_1, a_2, \ldots, a_{n_\gamma}) = a$,

where $\{f_\gamma^{n_\gamma} | \gamma \in \Gamma\}$ is the set of operations on the set of factors $\{a_\beta | \beta \in \mathrm{B}\}, f_\gamma^{n_\gamma}$ is n_γ-dimensional operation. There exists one to one correspondence between the class of all real systems and the class of all finite algebraic systems $I(S) = \langle \{a_\beta | \beta \in \mathrm{B}\} | \Omega_S = \{f_\gamma^{n_\gamma} | \gamma \in \Gamma\} \rangle$ of inner factors of systems.

The principle of hierarchy revealed in the Definition 4.1

Let's formalize the goal of a system. The goal of the system S can be specified by a single or n dimensional predicate Q in the following way

$$\bigwedge_{i \in I} P_i \Rightarrow Q$$

or

$$\bigwedge_{i \in I} P_i(S) \Rightarrow Q(A_1, \ldots, A_n)$$

where $A_1, \ldots, A_n$ belong to the domain of predicate Q. Numerical characteristics of a system S or structural characteristics of a system S can be changed but the resulting modified system S' may satisfied P which is the property of the integrity of the system S. That is $P(S) \Rightarrow P(S')$, and therefore, to perform almost all the same functions as the system S, and, ultimately, to achieve the goal Q. Thus the numerical change or structural change of the system S within certain limits does not violate its integrity. This property will be call quasi sustainable one.

Definition 4.13 The system S is quasi sustainable one in relation to integrity property P if there exists the system S' and the congruence $\equiv_P$ on the system S' such that factor-model $S'/\equiv_P \cong S$.

The formalization of the property of the system's integrity with respect to the property P makes it possible to determine the class of innovation systems with respect to the integrity property P.

To determine the class of innovation systems and to study it one should make following definitions. Now let's define the attribute characteristics of the system.

Definition 4.14 Let Q be a target of a system S. The set of predicates $\{P_\alpha | \alpha \in \Lambda\}$ is called internal attribute signs of a system S if the formula

$$\bigwedge_{\alpha \in \Lambda} P_\alpha \Rightarrow Q$$

is an identically true formula for S.

Definition 4.15 Let Q be a target of a system S. The set of predicates $\{P_\beta | \beta \in \mathrm{B}\}$ is called external attribute signs of a system S if the formula

$$Q \Rightarrow \bigwedge_{\beta \in B} P_\beta$$

is an identically true formula for S.

Definition 4.16 Let Q be a target of a system S. The set of predicates $\{P_\gamma | \gamma \in \Gamma\}$ is called attribute signs of a system S if the formula

$$Q \Leftrightarrow \bigwedge_{\gamma \in \Gamma} P_\gamma$$

is an identically true formula for S.

Definition 4.17 Let S be a system with integrity property P. Then S is called an innovative system with deciding integrity property P if S can be off line in every super system S' such that S is P-pure in S' , that is containing it in a way that is not distorted P-connections.

An innovative system *S* with integrity property *P* differs from a system with integrity property *P* higher (not lower than in super system containing it) implementation performance numeric innovative properties, or the fact that analogues of a system with property *P* does not exist. The main indicators of innovation are: newness, the degree or level of newness, consumer value, degree of implementation in practice, effectiveness, the presence of a single indicator of efficiency, the phenomenon of "flash" (a synergistic effect) characterizing the beginning of the autonomous work of the innovation system, the graph of all links of all innovations.

Let's remind the examples of innovative systems. These for example are:

IT-technologies, autonomous work in digital libraries, expert systems and so on.

References

1. Mogilevsky, V.D.: Methodology of systems. Economics, Moscow (1999). (in Russian)
2. Prigozhyn, I., Stengers, I.: Order from chaos. A new dialogue between man and nature. Progress, Moscow (1986). (in Russian)
3. Ivanov, A.E.: Genesis of synergetics. Electron. Sci. Pract. J. Mod. Sci. Res. Innovations, 9 (2013). LNCS Homepage: http://web.snauka.ru/issues/2013/09/26327
4. Knyazeva, E.N.: Synergetic-30 years old. Interview with G. Haken. Questions Philos. (3), 53–61 (2000) (in Russian)
5. Landau, L.D.: D. ter Haar (ed.) On the theory of phase transitions, reprinted in Collected Papers of L.D. Landau (Pergamon Press and Gordon & Breach Science Publishers, 1965), pp. 193–216; the original papers were published in Phys. Z. Sowjet. 11(26) (1937); 11(545) (1937)
6. Wiener, N.: Cybernetics or management and communication in living organisms and machines, 3d printing. The Technology Press, Wiley, New York, Hermann et CIE, Paris (1949)
7. Malt'sev, A.I.: Algebraic Systems. Nauka, Moscow (1970). (in Russian)
8. Serdyukova, N.A.: On generalizations of purities. Algebra Logic **30**(4), 432–456 (1991)
9. Serdyukova, N.A., Serdyukov, V.I., Uskov, A.V., Slepov, V.A., Heinemann, C.: Algebraic formalization of sustainability in smart university ranking system, In: Jain, L.C., Howlett, R.J., Uskov, V.L. (eds.) Smart Innovation, Systems and Technologies Book Series (SIST, vol. 75) International Conference on Smart Education and Smart E-Learning, pp. 459–474 (2017)
10. Fucks, L.: Infinite Abelian Groups, vol. 1, p. 335. Academic Press, New York and London (1970) and vol. 2, p. 416. Academic Press, New York and London (1973)
11. Dictionary of the Russian language four volumes. In: Ushakov, D.N. (ed.) State Publishing House of Foreign and National Dictionaries (1935–1940) (second edition was published in 1947–1948) (in Russian)
12. Cambridge English Dictionary. LNCS Homepage: https://dictionary.cambridge.org/
13. Serdyukova, N.A., Serdyukov, V.I.: The new scheme of a formalization of an expert system in teaching. In: ICEE/ICIT 2014 Proceedings, paper 032, Riga (2014)
14. Serdyukova, N.A., Serdyukov, V.I., Slepov, V.A.: Formalization of knowledge systems on the basis of system approach, SEEL2015, In: Jain, L.C., Howlett, R.J., Uskov, V.L. (eds.) Smart Education and Smart E-Learning, Smart Innovation, Systems and Technologies, vol. 41, pp. 371–380. Springer (2015)
15. Serdyukova, N.A., Serdyukov, V.I.: Modeling, simulations and optimization based on algebraic formalization of the system. In: 19th International Conference on Engineering Education July 20–24, 2015, Zagreb, Zadar (Croatia), New Technologies and Innovation in Education for Global Business, Proceedings, pp. 576–582. ICEE2015 (2015)

16. Serdyukova, N.A., Serdyukov, V.I., Uskov, A.V., Slepov, V.A., Heinemann, C.: Algebraic formalization of sustainability in smart university ranking system. In: Jain, L.C., Howlett, R.J., Uskov, V.L. (eds.) Smart Innovation, Systems and Technologies Book Series (SIST, vol. 75) International Conference on Smart Education and Smart E-Learning, pp. 459–474. Springer (2017)
17. Ershov, Y.: Profinite groups. Algebra Logic **19**(5), 552–565 (1980)

Chapter 5
Formalization of System Links: Different Approaches. Duality in Smart Systems Theory

Abstract In this chapter we consider the following main questions: different approaches to the definition of duality in Smart Systems Theory, measurement of the system's links strength, the group of links on the Cayley Graph of the system, the concept of efficiency and its formalization, the concept of *P*-efficiency of a system, *P*-subgroups of effective links of a system. In addition to Chap. 4 in this chapter we shall continue to study the links of the system and define another group of system's links as a group defined on the Cayley graph of the group of factors that determine the system. Then we proceed to study the subgroups of effective connections of the system. We shall also introduce the notion of a common efficiency problem with risks. The next question that we shall consider in this chapter is the use of duality in systems theory. We shall give a brief survey of some of the methods that are useful to investigate the duality.

Keywords Duality in smart systems theory · *P*-efficiency of a system
Group of system's links

5.1 Preliminary Facts

The question of duality, and, in particular, the question of duality in mathematics, is one of the most interesting questions connected, notably, with philosophy [1]. One of the methods for studying duality in systems theory is the tensor method of dual networks. The peculiarity of the use of the tensor method in the modeling and analysis of complex systems consists in the general method of modeling the processes of the system and the structure of the system, as well as in the calculation and analysis in this model changes of processes parameters when the structure is changed, including the decomposition of the system into subsystems and the synthesis of subsystems into a whole [1]. In [1], a classification of the system's relations was proposed. We note that the classification of connections, proposed in [1], fits into the model of hierarchy of structural connections of the system, constructed in Chap. 4.

N. Serdyukova and V. Serdyukov, *Algebraic Formalization of Smart Systems*,
Smart Innovation, Systems and Technologies 91,
https://doi.org/10.1007/978-3-319-77051-2_5

In fact, the classification according to [1] distinguishes the following types of connections:

(1) the relationships between the elements of a complex system,
(2) the relationships between the links of the system,
(3) the connection between the structure and the process- according to [1], energy, the concept of which is not formalized—the relationship between the elements of the system and the connection according to the classification constructed in Chap. 4—the hierarchy of structural links of the system.

Complementing the classification of structural links constructed in Chap. 4, by the links between the elements of the system and its first-level connections, we get a connection between the structure and the process according to [1]. According to our classification of the hierarchy of structural links, built in Chap. 4, these are the third level links. The model of the hierarchy of structural links of the system, constructed in Chap. 4, can be supplemented by a general model of relationships between the structure and processes of the system as follows: to do this, it is enough to define the links between $C_n(S)$ and $C_m(S)$, where n, m are natural numbers.

Also, questions of duality are studied in string theory [2, 3]. String Duality is the statement that one kind of string theory compactified on one space is equivalent in some sense to another string theory compactified on a second space. This draws a connection between two quite different spaces. Mirror symmetry is an example of this. The development of string theory stimulated the development of mathematical formalisms, mainly algebraic and differential geometry and topology. In this connection, we should note also that the geometry of Riemann surfaces plays an important role in string theory, and that papers on the algebraic geometry of algebraic systems have appeared recently [4]. It seems to us that the question of duality is closely connected with questions about the connections of the system. Another problem that we shall consider in this chapter is the question of the strength of the system's connections. This question in system theory is well known and has been considered in numerous works and in various fields of science, especially in physics and chemistry [5], techniques [1], and social sciences [6–8].

5.2 Several Examples

Let's consider the following examples.

One can single out:

1. **Connections in social systems**.

In the late 1940s, before the widespread use of the Internet, Granovetter and Freeman explored social networks, and introduced the notion of strong and weak social ties through the formal separation of two classes of interpersonal relationships by the frequency and duration of contacts. An example of strong ties are links with relatives and friends, the weak are the links between the neighbors, acquaintances, acquaintances of friends, formal contacts at work. Granovetter hypothesized that within social networks, weak connections are more important than strong ones [9]. Granovetter's conjecture is explained by the fact that information spreads faster and more widely through weak links. According to Granovetter, weak links are "necessary to expand the opportunities for users to interact with the community, whereas as a result of strong ties, a local link is formed." For example, people find work precisely because of weak connections, but not strong ones. Weak connections between people are characterized by the fact that we know little of these people, we do not see them constantly, but they are more useful in solving certain issues.

Granovetter showed that this is because, through strong ties, people share a limited amount of data or resources, while strong links are information-redundant, and thus less useful to each other [9].

2. **Connections in physical systems, string theory**.

Large and small scales, strong and weak links constants have been considered always as clear limits that determine the behavior of physical systems. String theory studies strong and weak connections, as well as duality, and can eliminate some differences between big and small, strong and weak, thanks to the use of Riemann surfaces [3]. Strong and weak links are distinguished by a link constant. The coupling constant is a number that characterizes the strength of the interaction, that is, how strong is the interaction. For example, the Newtonian gravitational constant is the link constant for the gravitational force.

Each force is characterized by its link constant. For example, in the case of electromagnetism, the link constant is proportional to the square of the electric charge. When studying the quantum behavior of electromagnetism, it did not work out to construct an exact theory describing behavior on all energy scales. Therefore, the whole range of energies was broken up into parts and a corresponding solution was built for each of the parts. Each of these ranges has its own link constant. Near "normal" energies the link constant is small, and in the next few ranges it can be used as a good approximation to its real values. However, in those ranges where the link constant is large, the methods used in working with "normal" energies do not run. A similar picture takes place in string theory. String theory also has its own link constant. However, unlike theories of elementary particles, string link constants are not just numbers, they depend on the vibrational modes of the string. The change in the field of the string mode to the same number, but with the opposite sign, changes the link constant from very large to very small. This type of symmetry is called *S*-duality. If two theories are related to each other by *S*-duality, then the theory which is taken with a strong link constant, will be equivalent to another theory, which is taken with a weak link constant.

3. **Connections in the tensor method of dual networks.**

The tensor in mathematics is an object which is characterized by an array of indicators as follows:

Let V be a an n—dimensional vector space, $n < \infty$. The conjugate or dual space V' to the space V is defined as the space of linear functions from V into the field of real numbers R. If the dimension of the vector space V is equal to n, then the dimension of the conjugate vector space V' is also equal to n. Vectors from V are called contravariant, and vectors from V' are called covariant. If to each coordinate system in the n—dimensional Euclidean vector space V there corresponds a system of n^{p+q} numbers $a_{ij\ldots}^{rs\ldots}$, where the number of lower indices is equal to p, and the number of superscripts is equal to q, in such a way that when passing from one coordinate system to another, these numbers are transformed by the formula

$$a'^{rs\ldots}_{ij\ldots} = c_i^{\propto} c_j^{\beta} \ldots b_{\sigma}^{r} b_{\tau}^{s} \ldots a_{\alpha\beta\ldots}^{\sigma\tau\ldots} \tag{5.1}$$

where $\|c_i^j\|$ is a matrix of the transition from one basis of the vector space V to another basis, and $\|b_i^j\|$ is a matrix which is transposed to the matrix inverse to the matrix $\|c_i^j\|$, then we say that a tensor is given which is p times covariant and q times contravariant. The number $p+q$ is called the rank or the valence of the tensor, and the numbers $a_{ij\ldots}^{rs\ldots}$ are called the tensor components [10]. So, a tensor of a type (p, q) is a correspondence between the bases of a finite–dimensional vector space V and the sets of n^{p+q} numbers $T = \left\{a_{ij\ldots}^{rs\ldots}\right\}$, for which the sets corresponding to different bases are connected by the relation (5.1). Tensor calculus makes it possible to formulate and to consider physical laws which remain in force when pass from one coordinate system to another. Tensors play an important role in differential geometry, quantum mechanics, fluid mechanics, elasticity theory, in the general theory of relativity. Particular cases of tensors are vectors and scalars.

The foundations of the tensor calculus were established in the works of K. Gauss (1777–1855), which were relating to the geometry of surfaces. Than G. Grassmann (1809–1877) expanded the theory of numbers by including tensor algebra in it, and B. Riemann (1826–1866), using Gaussian internal coordinates, considered n—dimensional manifolds in geometry. An important step towards the creation of a common tensor calculus was done by E. Christoffel (1829–1900) in the work on transformations of differential quadratic forms. In the 1890s, the Italian geometer G. Ricci-Curbastro (1853–1925) and his former student T. Levi-Civita (1873–1941) generalized and systematized the results of their predecessors.

So, we have that the tensor is an object of linear algebra that linearly transforms elements of one linear space into elements of another linear space. Besides it, the tensor is a mapping that allows one to concentrate huge information arrays. On the set of tensors, the operations of the tensor product, convolution and operation of indices lifting are introduced, which can be considered as a generalization of matrix operations, such as matrix multiplication, scalar product. Tensor operations follow from the multilinearness of tensors after the decomposition of vectors which are convolutions with tensors on the basis of a vector space, just as matrix operations follow from the linearity of linear operators and bilinear forms, each of which is represented by a concrete matrix in a particular basis. The tensors with the help of these operations, are associated with fundamental geometric objects, such as vectors and scalars, that determines tensors geometric meaning. These same operations connect tensors with matrices of coordinate transformations—Jacobi matrices. In [1], the tensor method of dual networks is used to calculate the changes in the processes when the structure of the elements of the system is changed, or, in a broad sense, to calculate the interaction of system processes and system structure.

Let us dwell on how one can determine the strength of the system's connections in the general case when using the algebraic formalization of the system.

5.3 System Connections Strength. Example: The Social Relationships Strength

Let's introduce the following definitions.

Definition 5.1 Let S be a system and $\boldsymbol{G_S}$ be a group of factors that determined the system S. The measure $PC(G_S)$ of the system S links strength is the number of possible different synergetic effects of the system S, that is the number of possible different final states of the system S, which are calculated by the model $\boldsymbol{G_S}$, or, which is the same, the number of pairwise nonisomorphic groups of order $|\boldsymbol{G_S}|$.

Notes

(1) Than $PC(G_S)$ index is larger, that the system S links calculated on the model $\boldsymbol{G_S}$ is weaker.
(2) Than $PC(G_S)$ index is smaller, that the system S links calculated on the model $\boldsymbol{G_S}$ is stronger.
(3) The system links indicator $PC(G_S)$ is a relative one and it depends on the choice of the model $\boldsymbol{G_S}$ of factors that determined the system S.

Definition 5.2 Let S be a system and $\boldsymbol{G_S}$ be a group of factors that determined the system S. Let $\emptyset \neq M \subseteq G_S$. The measure $PC(M)$ of the set M links strength is the number of possible different synergetic effects of the system $\langle G_S \backslash M \rangle$, where $\langle G_S \backslash M \rangle$ is a subgroup of the group $\boldsymbol{G_S}$, generated by the set $G_S \backslash M$.

Notes

(1) Than $PC(M)$ index is larger, that the set M links calculated on the model $\boldsymbol{G_S}$ is weaker.
(2) Than $PC(M)$ index is smaller, that the set M links calculated on the model $\boldsymbol{G_S}$ is stronger.

(3) The system links indicator $PC(M)$ is a relative one and it depends on the choice of the model $\boldsymbol{G}_S$ of factors that determined the system S.

The links strength indicators introduced in Definitions 5.1 and 5.2 explain Granovetter's theory.

5.4 Duality in System Theory

In this chapter other versions of constructing a duality theory for the theory of systems are proposed. An important role in this matter is played by the formalization of the concept of the connection of the system and the clarification of its meaning. Here several ways of formalizing the links of the system are proposed.

The first way to construct a formalization of the system's links runs as follows. The visual representation of the connections of the system uses graph theory. We have constructed on this basis a group of the system's links that uses the Cayley graph of the group of factors G_S, determining the system S and the construction of the free product.

The second way runs as follows. Let the system link connects some elements a, b of the system and we are examine the model of factors which determine the system S. Let this model be an algebraic system $\boldsymbol{A}_S = \langle A_S, \Omega \rangle$ of the signature Ω. The system's links should preserve but not destroy the internal structure of the system. So, it is natural to consider the homomorphisms of the system $\boldsymbol{A}_S$ into itself, that is the maps of the set A_S into itself, preserving operations and predicates from Ω, as the system's links.

Hence from we obtain several ways to study duality in smart systems theory.

The first way of constructing duality for the theory of smart systems uses models of factors that determine the system, and these models of factors are algebraic systems $\boldsymbol{A}_S = \langle A_S, \Omega \rangle$ of some signature Ω. Further the classical theory of duality from category theory is used in this method [11, 12]. It follows from the existence for each category the dual one that there works a duality principle in the category theory, that is, for every true sentence of the predicate calculus with respect to one category there exists a dual true statement for the dual category. The statement Pr^D, which is dual to the statement Pr, and is formulated in the language of category theory, is obtained by interpreting in the category $\Re$ the sentence Pr, considered in the dual category $\Re^D$. A dual statement is obtained from the original one by preserving the logical structure of the statement and replacing in its formulation all the arrows by the opposite, and all products of morphisms into products of morphisms written in the reverse order.

The second method was proposed by us for the case when the model of factors is a group of factors G_S. Here we can consider the following two cases.

The first case. The group of factors which determined the system S is finite, and $|G_S| = n$. It is well known that in this case the group $\boldsymbol{G}_S$ can be embedded in the symmetric group of all permutations $\boldsymbol{S}_n$ of degree n. The second case does not use restrictions on the number of elements of the group $\boldsymbol{G_S}$. In the second method, we propose to embed $\boldsymbol{G_S}$ in its holomorph $Hol\boldsymbol{G_S}$. First of all let's consider the case where $\boldsymbol{G_S}$ is a finite abelian group. Then the holomorph $Hol\boldsymbol{G_S}$ of the group $\boldsymbol{G_S}$ is a semidirect extension of the group $\boldsymbol{G_S}$ with the help of its group of automorphisms $Aut(\boldsymbol{G_S})$. Let's use the multiplication form of the record for a group operation in $Aut(\boldsymbol{G_S})$, and for a group operations in $\boldsymbol{G_S}$ and in $Hol\boldsymbol{G_S}$ let's use $\circ$ and $+$ respectively. The main set of the group $Hol\boldsymbol{G_S}$ can be considered as the set of all ordered pairs (g, φ), where $g \in \boldsymbol{G_S}$, $\varphi \in Aut(\boldsymbol{G_S})$. The group operation is given in $Hol\boldsymbol{G_S}$ by the rule: $(g, \varphi) + (h, \psi) = \left(g \overset{\circ}{} \varphi h, \varphi\psi\right)$ for any $(g, \varphi) \in Hol\boldsymbol{G_S}$, $(h, \psi) \in Hol\boldsymbol{G_S}$.

In general holomorph of a group is the concept of Group Theory that arose in connection with the solution of the following problem: is it possible to include any given group $\boldsymbol{G}$ as a normal subgroup in some other group so that all automorphisms of $\boldsymbol{G}$ are consequences of inner automorphisms of this larger group? To solve this problem, we construct a new group $Hol(\boldsymbol{G})$, with respect to the group $\boldsymbol{G}$ and its automorphism group $Aut(\boldsymbol{G})$, whose elements are the pairs (g, φ), where $g \in \boldsymbol{G}$, $\varphi \in Aut(\boldsymbol{G})$, and in which the composition pairs according to the following formula:

$$(g_1, \varphi_1)(g_2, \varphi_2) = \left(g_1 \overset{\circ}{} \varphi_1^{-1}(g_2), \varphi_1\varphi_2\right)$$

Herewith the automorphisms $Aut(\boldsymbol{G_S})$ of the group of factors of the system $\boldsymbol{S}$ are in fact the links of the system $\boldsymbol{S}$ with special properties:

(1) the one-to-one correspondence between the factors that determine the system,
(2) the preservation by the link of the composition operation of the factors which determine the system.

In this connection, a special role here belongs to perfect groups, that is such groups $\boldsymbol{G}$ which are isomorphic to the group of its automorphisms $Aut(\boldsymbol{G})$. For example, $\boldsymbol{G} \cong \boldsymbol{S_n}$, where $n \neq 2, 6$. We have $Hol\boldsymbol{G}/\boldsymbol{G} \cong AutG \cong G$ for a perfect group $\boldsymbol{G}$. We obtain the following conclusion from all the above.

The Main Conclusion About Duality

The main idea of the representation of duality in the theory of systems runs as follows. Let $\{S_\propto | \propto \in \Lambda\}$ be a non- empty set of systems; and the element $a \in S_\propto, \propto \in \Lambda \Leftrightarrow \propto \in \bigcap_{\propto \in \Lambda} S_\propto \neq \emptyset$. All such elements a give us in point of fact the set of all connections between systems $S_\propto, \propto \in \Lambda$. Elements of a new system S^d which is dual to the system S are the systems $S_\propto, \propto \in \Lambda$ that is the elements of the set $\{S_\propto | \propto \in \Lambda\}$.

5.5 The Connection Between Duality and the Concept of a Factor of a System

Let's return to Example 3 from Sect. 5.2, and namely to the definition of the tensor and the conjugate or dual vector space. According to the definition of the conjugate or dual vector space, we have $V' = \{f | f : V \to R\}$, where every f is a linear function from the vector space V into the field of real number R. Let's consider the correspondence

$$V' = \{f | f : V \to R\} \mapsto \{Imf \cong V/kerf \leq R\}$$

This correspondence, as well as the use of the concept of the group holomorph, helps us to introduce the concept of a factor-system as a concept dual to the concept of a subsystem. This can be done for the model of algebraic formalization of the system by using the concept of factors which determine the system, and by using the concept of elements of the system for the system itself. In fact, let S be a system and $\boldsymbol{G_S}$ be a group of factors which determine the system S. Let's define an equivalence relation ρ on the set of factors G_S, that is ρ is a reflexive, symmetric and transitive relation. It means that $\rho \in G_S \times G_S$ satisfies the following conditions:

(1) $a\rho a$, or, that is the same $\langle a, a\rangle \in G_S \times G_S$ for every $a \in G_S$,
(2) $a\rho b \Rightarrow b\rho a$ for every $a, b \in G_S$,
(3) $a\rho b \Lambda b\rho c \Rightarrow a\rho c$ for every $a, b, c \in G_S$.

Besides it, we shall assume that ρ is a congruence on the group $\boldsymbol{G_S}$, that is ρ preserves the operation of composition and a feedback.

Definition 5.3 The factor-group $\boldsymbol{G_S}/\rho$ is called the model of the factor—system of the system S for the congruence relation ρ on the set of factors which determine the system S.

Examples

(1) ρ is an equality relation on the set of factors which determine the system,
(2) ρ is the relation of the interchangeability of factors which determine the system,
(3) ρ the relation of the equality of risk factors, that is, the measure of the risk factors that are connected by relation ρ, is quite the same,
(4) ρ is the relation of the equality on some fixed group of indicators defined on the set of factors G_S, which determines the system S.

Let $M = \{m_i | i \in I\}$ be a set of numerical indicators defined on the group of factors $\boldsymbol{G_S}$, that is every m_i is a map from the set G_S into the set R of real numbers: $m_i : G_S \to R$, where $i \in I$.

For $g, h \in G_S$ let $g\rho_i h$ be true if and only if $m_i(g) = m_i(h)$, $i \in I$. Then ρ_i is an equivalence on the set G_S for every $i \in I$. So, $\bigcap_{i \in I} \rho_i$ is an equality relation on the set of the indicators $M = \{m_i | i \in I\}$ on the group $\boldsymbol{G_S}$, where ρ_i is an equality relation by the indicator m_i. This raises the main question in studying the links of the model

$\boldsymbol{G_S}/\rho$. Let γ be a binary relation on the set G_S, which corresponds to the graph of links of the system S by the model $\boldsymbol{G_S}$, that is $\gamma \subseteq G_S \times G_S$ and $\gamma = \{g \in G_S | \exists h \in G_S : g\gamma h \, or \, h\gamma g\}$. Let's consider $\boldsymbol{N_\gamma(S)} = \langle g \in G_S | \exists h \in G_S : g\gamma \cap \rho h \, or \, h\gamma \cap \rho g \rangle$- the normal subgroup of the group $\boldsymbol{G_S}$, which is generated by the set $\{g \in G_S | \exists h \in G_S : g\gamma \cap \rho h \, or \, h\gamma \cap \rho g\}$. Then factor-group $\boldsymbol{G_S}/\boldsymbol{N_\gamma(S)}$ is a factor-group of the model of the system S, free from links up to indicator ρ (besides factors which are connected with ρ).

Now let's consider the notion of factor-set of the set of elements of the system S. Note that in such formulation of the problem, the requirement that τ should be a congruence on the set S becomes unnecessary. So, we should distinguished between the notions of a model for factor-system of the system S and the notion of a factor set of a system S.

Definition 5.4 Let S be a system and τ be an equivalence on the set of elements of the system S. Factor-set S/τ is called a factor-set of the system S.

Example

(1) Let τ be the relation of the interchangeability of elements on the set of elements of the system S. Then the factor system S/τ has no interchangeability property.

Once more question which appears here is the question about the links between elements of the factor-set S/τ. This question is solved as follows. For $a, b \in S$ let $a\rho b \Leftrightarrow$ there exists a link between a, b in the system S. So we can view instead of the factor-set S/τ the factor-model $\langle S, \rho \rangle/\tau$ of the model $\langle S, \rho \rangle$ by the relation of equivalence τ.

Definition 5.4 allows us to consider the elements of the system S as free from links, except for the link τ (with the exception of the elements of linked by τ).

Theorem 5.5 *Let S be a system and $\boldsymbol{G_S}$ is a group of factors which determine the system S. If $\boldsymbol{G_S}$ is a perfect group, then the power of links of the system S is no more than the power of links of the system of the first level dual to the system S.*

Proof The assertion of the theorem follows from the classification of the structural connections of the system constructed in Chap. 4, Definition 4.1 and the fact that for the perfect group $\boldsymbol{G_S} \cong Aut\boldsymbol{G_S}$.

Theorem 5.6 *Let S be a system and $\boldsymbol{G_S} \cong \boldsymbol{V}$, where $\boldsymbol{V}$ is an additive group of the Euclidean vector space of the dimension n, be a group of factors which determine the system S. Then the powers of links of the system S and the system S' dual to S and defined by the group of factors $\boldsymbol{V}'$, where $\boldsymbol{V}'$ is an additive group of the vector space V' which is conjugate or dual to the vector space V, are the same.*

Theorem 5.7 *Let S be a system and $\boldsymbol{G_S}$ be a group of factors which determine the system S. Let the order of the group $\boldsymbol{G_S}$ be equal to the product of two non-equal prime numbers: $|\boldsymbol{G_S}| = pq, p \neq q$. Then the power of links of every factor-system of the system S is less or equal to the power of links of the system S, defined by the model $\boldsymbol{G_S}$.*

The proof of the Theorem 5.7 follows from Lagrange's theorem about the order of a subgroup of a finite group.

5.6 Algebraic Formalization of Modeling the Processes Preserving the Operation of Composition of Factors of a Closed System

Definition 5.8 By the process of the system we mean the change in the elements and connections of the system during system functioning.

As a rule we use systems of equations-algebraic, differential, etc., connecting the known values of the numerical indicators of the process and the sought values of the numerical indicators of the process, which make it possible to characterize the process under study when modeling processes by classical methods. The question of modeling processes occurring in the system by the method of algebraic formalization arises, that is, by the method of translating questions of the general theory of systems into the language of the theory of algebraic systems, developed by A. I. Malt'sev We shall do the same in the solution of this problem, but we shall make several changes into this scheme.

Definition 5.9 Let S be a system and $\boldsymbol{G_S}$ be a group of factors which determine the system S, $T \subseteq R$ be a time interval for which processes occurring in the system S are examined. The group $Hom(\boldsymbol{G_S G_S})$, of homomorphisms of the group $\boldsymbol{G_S}$ is called the set of states of processes occurring in the system S. A process which takes place in a system S is a map $f_S : T \to Hom(\boldsymbol{G_S}, \boldsymbol{G_S})$.

Definition 5.9 makes sense, as the processes occurring in the system, and preserving the operation of composition of factors, are characterized by a change in the links between the factors that determine the system, that is, by the maps of the sets of links of the system into itself.

5.6.1 *Example. Modeling Decomposition Process of the System*

Decomposition of a complex system is the analysis of this system through its decomposition or separation into parts.

Decomposition, as a process of separation, allows us to consider any system as complex, consisting of separate interconnected subsystems, which, in turn, can also be divided into parts. Herewith, not only material objects, but also processes, phenomena and concepts can act as systems. Usually, while decomposing the system, the following algorithm is used. The investigated complex system is located at the zero level. After its decomposition, the subsystems of the first level are obtained. In turn, the decomposition of these subsystems or some of them leads to the appearance of subsystems of the second level, and so on. A process representation in the form of a graph is used for a simplified graphical representation of the decomposition process. Since the decomposition process according to the above algorithm is of a hierarchical nature, it can be represented as a graph, which is called a tree. A tree is a graph without closed paths, with vertices at certain levels. The equivalent definition of a tree graph runs as follows: A tree is a connected graph which contains no cycles. The representation of the decomposition process in the form of a graph allows one to visualize the decomposition process. We shall assume that the system links preserve the operation of composition of the factors which determine the system S, that is, the group operation $\boldsymbol{G_S}$, and also maintains a feedback for the factors which determine the system S, that is, the operation of taking the inverse element in the group $\boldsymbol{G_S}$. Hence it turns out that the connections of the system S in the model $\boldsymbol{G_S}$ can be interpreted as homomorphisms of the group $\boldsymbol{G_S}$ into itself. Let $f : \boldsymbol{G_S} \to \boldsymbol{G_S}$. Since $\boldsymbol{Imf} \cong \boldsymbol{G_S}/\boldsymbol{kerf}$, one finds out that the study of the decomposition process of the system S is related with the study of normal subgroups of the group $\boldsymbol{G_S}$ and the factor-groups of the group $\boldsymbol{G_S}$, and the group $Hom(\boldsymbol{G_S}, \boldsymbol{G_S})$. The homomorphism f can be interpreted as a link of the system $\boldsymbol{G_S}$, which is nullified, that is not acting, on the elements of the subsystem $\boldsymbol{kerf} \trianglelefteq \boldsymbol{G_S}$. Factor-group $\boldsymbol{Imf} \cong \boldsymbol{G_S}/\boldsymbol{kerf}$ shows the scope of link $f : \boldsymbol{G_S} \to \boldsymbol{G_S}$ realization. Thus, the homomorphism group $Hom(\boldsymbol{G_S}, \boldsymbol{G_S})$ represents the connections of the system S in the case when S is represented by the group $\boldsymbol{G_S}$. We shall assume that the decomposition of a complex system is the allocation of all subsystems of the system with preservation, if possible, all of the initial links of a complex system, that is, decomposition is a step-by-step description of the structure of the system in a section with including in this description the structure of the subsystems of the original complex system. So we get that if the system S is represented by a group of factors $\boldsymbol{G_S}$, defined the system S, then the split-off in S a subsystem in which the link f does not function or the decomposition of the system S by the link f in the model $\boldsymbol{G_S}$ is represented with the help of the exact sequence

$$E \to \boldsymbol{G_i} \xrightarrow{f} \boldsymbol{G_S} \xrightarrow{\pi} \boldsymbol{Imf} \cong \boldsymbol{G_S}/\boldsymbol{kerf} \to \boldsymbol{E} \tag{5.2}$$

where $\boldsymbol{G_S}/\boldsymbol{kerf}$ shows the scope of link $f : \boldsymbol{G_S} \to \boldsymbol{G_S}$ realization. If the extension (5.2) splits, that is there exists such homomorphism $\sigma : Imf \to \boldsymbol{G_S}$, that $\pi\sigma = 0$ then the decomposition of the system S over the link f in the model $\boldsymbol{G_S}$ is a semidirect product

$$\boldsymbol{G_S} = \boldsymbol{kerf} \leftthreetimes \boldsymbol{Imf}$$

Here $\boldsymbol{Imf}$ is a complement subgroup or a complement. Let's consider the process of decomposition of the system S into closed subsystems $\{\boldsymbol{G_i} | i \in I\}$ by the links of the system S on the model $\boldsymbol{G_S}$.

As $Hom(\mathbf{G}_i, \mathbf{G}_S) \leq Hom(\mathbf{G}_S, \mathbf{G}_S)$, then $h_S : T \to Hom(\mathbf{G}_i, \mathbf{G}_S) \leq Hom(\mathbf{G}_S, \mathbf{G}_S)$, where $T \subseteq R$, describes the process of decomposition of the system S onto closed subsystems $\{\mathbf{G}_i | i \in I\}$ in time.

Let's consider the following example.

Example. Decomposition of the Education System by Goals and Links

We shall continue to consider the decomposition of the education system, which was started in Chap. 4 by examining the decomposition of THE World University Rankings. We can single out the following subsystems in the education system:

1. Subsystem of knowledge S_1 (an informational subsystem of the system S)
2. Subsystem of pedagogy or methodological and methodical subsystem S_2 (an adaptive subsystem of the system S)
3. Subsystem of students S_3 (a target subsystem of the system S)
4. A financial subsystem S_4 (a providing subsystem of a system S).

   According to this decomposition one can single out the following goals which are correspond to the distinguished subsystems.

5.7 Duality in the Theory of Strong and Weak System's Links

The Theorem 5.6 that we have proved explains the role of duality in the theory of strong and weak links of the system.

Theorem 5.6 *Let S be a system and $\mathbf{G}_S \cong \mathbf{V}$, where $\mathbf{V}$ is an additive group of the Euclidean vector space of the dimension n, be a group of factors which determine the system S. Then the powers of links of the system S and the system S' dual to S and defined by the group of factors $\mathbf{V}'$, where $\mathbf{V}'$ is an additive group of the vector space V' which is conjugate or dual to the vector space V, are the same.*

5.8 Efficiency (Utility of a Smart System). Formalization of Efficiency

The concept of efficiency is the basic one in the science of Operations Research. The development of Operations Research as a science is usually associated with the work of L.V. Kantorovich "Mathematical methods of organization and planning of production," 1939 [13], and also with the fortieth years of the twentieth century. The second starting point is the work of 1947 by J. Dantzig, devoted to the solution of linear extreme problems [14]. However, currently there is no rigid, well-established and generally accepted definition of the subject of Operations Research. It is generally accepted that Operations Research is a set of scientific methods for solving problems in developing recommendations for the effective management of organizational systems. It is generally accepted that Operations Research is a set of scientific methods for solving problems in developing recommendations for the effective management of organizational systems. The nature of organizational systems can be very different, but they have common mathematical models that are used to solve production, economic problems, in biology, in sociological research and other practical fields. The very name of the discipline is associated with the use of mathematical methods for the management of military operations.

Let us recall the basic concepts of Operations Research. One of the key concepts of Operations Research is the concept of efficiency. It is believed that the term "efficiency" appeared for the first time in the economic literature.

The foundations of the doctrine of economic efficiency are set forth in the work of the Italian economist Pareto [15]. An effective state of the Pareto system is a state of the system in which it is impossible to improve the state of any of its elements without worsening the state of its other elements. To compare the states of the system, the efficiency criteria are used, according to which the comparison takes place. The criterion is a rule, according to which the possible states of the controlled system are compared, achievable with the help of certain operations. It follows from this that efficiency is a relative concept.

Formalization of the criterion of effectiveness is called an indicator of effectiveness.

Definition 5.10 An operation aimed at achieving state A is a system of actions which goal is to achieve state A.

One of the first representatives of management theorists G. Emerson, singled out the relationship between efficiency and functionality [16]. This relationship allows us to determine the effectiveness as a utility function.

Definition 5.11 The efficiency function $u : X \to R$, defined on the set of all final states of the system S can be defined as a utility function with properties: for any $x, y \in X$, if $x \succsim y$, then $u(x) \geq u(y)$.

Let us explain this definition and give some examples of a possible definition of the efficiency function of the system S.

Examples of setting up the smart system efficiency function.

1. The effectiveness of the system by the final states of the system. In the first example G_S is a group of factors which determined the system S, $X = \{G_i | i \in I\}$ is a set of all final states of the system S, S and G_S are finite, so X is finite. The relation $\succsim$ on the set X reflects preferences about the finite state of a system.
2. The effectiveness of the system's links according to the levels of the system's connections. We set up $u(G_k(S)) \geq u(G_m(S)) \Leftrightarrow k \geq m$, where $G_k(S)$ and $G_m(S)$ are the links of the level k and m the system S respectively.
3. Efficiency (usefulness) of the system by the links of the system of the same level. Let $a, b \in G_k(S)$ are the links of the level k of the system S, $\propto_k$ is a relation of link preference at the level k. Then $u(a) \geq u(b) \Leftrightarrow a \propto_k b$.

4. Efficiency (utility) of the system by the group of links of the system $\Gamma = *_\alpha a_\alpha H$ by Cayley table of a group of factors графа G_S, where the group of links of the system is defined as follows.

Definition 5.12 Let $G(S) = \langle G, \cdot, \square^{-1} \rangle$ be a group of factors that represents system S, and $G_S = \{a, b, \ldots\}$ be a main set of group $G(S)$. Let $\Gamma'_S = \{v_{ab} | a, b \in G_S\}$ be the Cayley's graph for $G(S)$, or, in other words, there exists an edge v_{ab} if and only if $b = a * h, h \in H \cup H^{-1}$, where H is a set of generators of a group $G(S)$. Let $\Gamma_S = \{v_{aa} | a \in G_S\} \bigcup \Gamma'_S$ Let $a, b \in G_S$ and $V_{ab} = \{v^h_{ab} | a, \quad b \in G_S \,\&\, \exists\ h \in H^{-1} \cup H : b = ah\}$. In this case, $V = \bigcup_{a \in G_S} V_{ab}$ is a set of all links between elements $, b \in G_S$such that $\exists h \in H^{-1} \cup H : b = ah$.

Let $M = V \cup \{v_{aa} a \in G_S\}$. Let $G_a = \langle ah | h \in H^{-1} \cup H \rangle$ be a subgroup of $G(S)$ generated by the set $\{ah | h \in H^{-1} \cup H\}$. In this case, $G_a \cap G_b \neq E \overset{\square}{\Leftrightarrow} a^{-1} b = hk^{-1}, h, k \in H^{-1} \cup H$, where E is a unit subgroup of $G(S)$.

Let us consider a free product $F_S = \underset{a}{*} G_a$, where a free product is taken over all $a \in G_S$ in such a way that for every free factors G_a and G_b, where $a, b \in G_S$ we will get has $G_a \cap G_b = E$. In this case, F_S is called a group of links of system S. (A note: a free product operation defines all possible best combinations of links).

The notion of the group of links of the system is important by the following considerations.

Let's consider the decomposition of a system S at elements: $S = \coprod_i S_i$, where every S_i is a subsystem of a system S or an atomic element of a system S. Usually, the flat graphs are used to represent links of a system. A flat graph is a picture on a plane that consists of a set of points $\{a\}$ (which are the vertices of a graph with one to one corresponding to a set $\{a = S_i\}$) and a set of edges of a graph (which are directed arks v_{ab}) if there exists a link between $a = S_i$ and $b = S_j$ in the system S. If a link is two-dimensional, then arc v_{ab} is non-direct one. There are a lot of examples which show that it is necessary to use multigraphs (or pseudo-graphs) – graphs in which multiple edges are allowed, i.e. edges with the same vertices during a simulation. For example, modeling a possible trajectory of an aircraft requires a utilization of a multigraph. However, a multigraph, which represents a system, does not provide information to identify quality of system's links. Various methods, that are used in mathematical statistics and econometrics, are able to evaluate quantitative characteristics of system's links quite strong constrains. However, they are useless to analyze system's qualitative characteristics.

5.9 Presentation of the General Task of the Smart System Effectiveness Determining in the Form of an Optimization Problem with Risks

The next question that arises in the study of the utility function or the effectiveness of the system is the question of how to link the effectiveness function and the effectiveness criterion with performance indicators. The numerical or formalized expression of an efficiency criterion is a system of equations or inequalities that denotes the limits in which an efficiency index must be found, while the utility function, depending on the meaning of the problem, has the form: $u \to max$ or $u \to min$. Thus, we have the usual form of a problem for a maximum or a minimum:

$$\begin{cases} u(x_1, x_2, \ldots, x_m) \to \max(\min) \\ w_1(x_1, x_2, \ldots, x_m) = 0(\geq 0) \\ \ldots\ldots\ldots\ldots\ldots\ldots\ldots\ldots\ldots\ldots\ldots \\ w_n(x_1, x_2, \ldots, x_m) = 0(\geq 0) \end{cases} \tag{5.3}$$

Further, this task can be modified by going to the task for a maximum or a minimum with risks. To do this, we should select the permissible risks for the left parts of the system constraints

$$\begin{cases} r_1 \leq w_1(x_1, x_2, \ldots, x_m) \leq R_1 \\ \ldots\ldots\ldots\ldots\ldots\ldots\ldots\ldots\ldots\ldots\ldots \\ r_n \leq w_n(x_1, x_2, \ldots, x_m) \leq R_n \end{cases} \tag{5.4}$$

or variables $x_1, x_2, \ldots, x_m$:

$$\begin{cases} r_1 \leq x_1 \leq R_1 \\ \ldots\ldots\ldots\ldots\ldots\ldots\ldots\ldots\ldots\ldots \\ r_m \leq x_m \leq R_m \end{cases} \tag{5.5}$$

If, herewith, we can find a linear dependence for the efficiency function $u(x_1, x_2 \ldots x_m)$, expressing its dependence of the parameters or indicators $x_1, x_2, \ldots, x_m$, then we obtain a linear programming problem with risks:

$$\begin{cases} u(x_1, x_2, \ldots x_m) \to \max(\min) \\ r_1 \leq x_1 \leq R_1 \\ \ldots\ldots\ldots\ldots\ldots\ldots\ldots\ldots\ldots\ldots \\ r_m \leq x_m \leq R_m \end{cases} \tag{5.6}$$

5.10 Examples. The Use of Duality for Complex Smart Systems Classification by the Number of System Goals. Stability by the Parameter of Achieving the Goal of the System

The main conclusion about the representation of duality in the theory of systems allows us to tie up the proposed constructions to the classical approach of describing the properties of the system, which basically uses the notion of an element of the system, and not the factor which determines the system. This approach allows, for example, to classify complex smart systems S according to the number of goals of system S as follows to study the stability of smart systems in terms of the parameter of achieving the goals of the functioning of the system. A complex system is a multipurpose system designed to achieve either one, or another, or the third, etc. goals. At the same time, it is not known beforehand in what sequence the system will have to solve the tasks that are to achieve one or another goal. Presumably, in a complex system, it is possible to single out a subsystem designed to achieve one of these goals. At the same time, there may appear "extra" elements of the system or even "extra" subsystems of the system that are not participating in achieving this goal, but will be involved in achieving another goal. Let us explain this by an example. Let's suppose that there are three goals to achieve which a smart system is created. Let's suppose that there are three goals to achieve which a smart system is created. Then the following classification of subsystems and elements of the initial smart system arises.

- a set of subsystems and elements, as well as their relationships that will be involved in achieving any of the three goals of the system, if necessary,
- the totality of the elements, as well as their links, which will be involved in solving problems when they achieve:
- the first or second goal,
- the first or third goal,
- the second or third goal,
- a set of subsystems and elements, as well as their links, which will be used to solve problems when they achieve:
- the first goal,
- the second goal,
- the third goal.

Next, we single out the core of a complex smart system - a set of subsystems and elements, as well as their links, which will be used to solve problems when all the goals of the smart system should be achieved. This classification can be formalized and generalized to any finite number of goals of the system.

Let S be a system and $\{A_1, \ldots, A_n\}$ are the goals of the system S functioning. Let $S_k = \{a_{k1}, \ldots, a_{kn_k}\}$ be an aggregate of elements of the system S, involved in achieving the goal $A_k, k = 1, \ldots, n$.

$F_k = \{f_{k1}, \ldots, f_{km_k}\}$ be the links between the elements of the set S_k, which will be needed when solving problems on the achievement of the goal $A_k, k = 1, \ldots, n$. Let us define the relations $\rho_k, k = 1, \ldots, n$ on the set of the elements of the system S as follows.

$a\rho_k b \Leftrightarrow a, b \in S_k$. Then ρ_k is an equivalence on the set of elements of the system S, and we can examine the factor-system of elements S/ρ_k. Factor –system S/ρ_k would not reach the goal $A_k, k = 1, \ldots, n$. The intersection $\bigcap_{k=1,\ldots,n} \rho_k$ is a core of the system S. Factor-system $S/\bigcap_{k=1,\ldots,n} \rho_k$ would reach no one of the goals $A_1, \ldots, A_n$.

References

1. Petrov, A.E.: Tensor method of dual networks. LLC “Center for Information Technologies in Environmental Management”, Moscow (2007)
2. Aspinwall, P.S.: String theory and duality. In: Proceedings of the ICM. vol. 2. pp. 229–238. DUKE-CGTP-98-06 math.AG/9809004 (1998)
3. Yau, S.T., Nadis, S.: Sting Theory and the Geometry of the Hidden Dimensions, Basic Books. A Member of the Perseus Books Group, New York (2010)
4. Daniyarova, E.Y., Myasnikov, A.G., Remeslennikov, V.N.: Algebraic geometry over algebraic systems. II. Found. Fundam. Appl. Math. 17(1), 65–106 (2011/2012)
5. Rebinder, P.A.: Selected works. Surface phenomena in disperse systems. In: Colloid Chemistry. Science, Moscow (1978)
6. Granovetter, M.S.: Getting a Job: A Study of Contacts and Careers, 2nd edn. University of Chicago Press, (1985) (orig 1974)
7. Freeman, L.C, White D.R., Romney, A.K. (eds.): Research Methods in Social Network Analysis. George Mason University Press, Fairfax, VA (1989)
8. Burt, R.S.: Structural Holes: The Social Structure of Competition. Harvard University Press, Cambridge (1992)
9. Granovetter, M.S.: The strength of weak ties. Am. J. Psychol. **78**(6), 1360–1380 (1973)
10. Gelfand, I.M.: Lectures on linear algebra. Interscience Publishers, Inc., New York, LTD, London (1998)
11. Bucur, I., Deleanu, A.: Introduction to the theory of categories and functors, pure and applied mathematics, vol. 19. Interscience Publishers (1968)
12. Artamonov, V.A., Salii, V.N., Skornyakov, L.A., Shevrin, L.N., Shul’geifer, E.G.: General Algebra, vol. 2. Science, Main Edition of the Physical and Mathematical literature, Moscow (1991). (in Russian)
13. Kantorovich, L.V.: Mathematical methods of organization and planning of production. Manage. Sci. **6**(4), 366–422 (1960). (origin 1939 in Russian)

14. Dantzig, G.B.: Reminiscences about the origins of linear programming. Oper. Res. Lett. **1**, 43–48. Also in Bachem, A., Grotschel, M., Korte, B. (eds.) Mathematical Programming: The State of the Art, Bonn 1982, pp. 78–86. Springer, Berlin (1983); also in Memoirs of the American Mathematical Society vol. 48, pp. 1–11 (1984). Also in Cottle, R.W., Kelmanson, M.L., Korte, B. (eds.) Mathematical programming, Proceedings of the International Congress on Mathematical Programming, Rio de Janeiro, Brazil, April 6–8, 1981, North-Holland Publishing Co., Amsterdam, 1984, pp. 105–112
15. Pareto, V.: Manuel of political economy, a critical and variorum edition. In: Montesano, A., Zanni, A., Bruni, L., Chipman, J. S., McLure, M. (eds.) (2014) (origin 1906)
16. Emerson, H.:The twelve principles of productivity. Economics (1992). (origin The Engineering Magazine Co. New York (1919))

Chapter 6
P-Innovative and *P*-Pseudo-Innovative Systems on the Predicate *P* and Their Properties

Abstract In this chapter we shall continue to study the concept of efficiency, and in line with this concept we introduce the definition of an innovative smart system. After that in terms of algebraic formalization we shall describe the structure of innovative smart systems. Then we shall consider the concept of pseudo-innovation system, dual to the concept of an innovative smart system. Also an algebraic formalization of some properties of innovative smart systems and pseudo-innovative systems is constructed. Some examples of the use of these concepts consumed in the expert systems in training, and in the economy are given. After that we shall continue the empirical study of the process of system's decomposition using the example of the decomposition of the education system on the basis of these questions. The algorithm for a comprehensive assessment of the effectiveness of the functioning of the innovation system based on the tensor estimation of system performance is also proposed in this chapter. It is proposed to use homomorphisms of the group $\boldsymbol{G}_s$ of factors defining the system S into the group $GL(n,\Delta)$ of linear homogeneous transformations of the vector space R^n as tensor estimates of the efficiency of the functioning of the system S. One can also consider homomorphisms of the group of factors $\boldsymbol{G}_s$ that define a system S in the group $GL\ (n,\Delta)$ of linear homogeneous transformations of the vector space Δn over an arbitrary field Δ.

Keywords Innovation · Effectiveness · Tensor estimation

As we mentioned before, in this chapter we shall consider the following questions. We shall continue to study the concept of efficiency, the structure of innovative smart systems. To study the structure of the innovation system, the algebraic formalization of this concept will be used. In addition, we shall trace the analogy between the concept of an innovative smart system and the concept of an inverse limit, as well as analogues of the theorem on the description of abelian algebraically compact groups [1]. Then concept of pseudo-innovation system, dual to the concept of an innovative smart system, will be considered and we shall get a description of pseudo-innovative systems using algebraic formalization. The notion of a

N. Serdyukova and V. Serdyukov, *Algebraic Formalization of Smart Systems*, Smart Innovation, Systems and Technologies 91,
https://doi.org/10.1007/978-3-319-77051-2_6

quasi-sustainable system is introduced by the analogy with the concept of quasi-isomorphism from the Abelian Groups Theory. An algebraic formalization of some properties of innovative smart systems and pseudo-innovative systems is constructed. Some examples of the use of these concepts consumed in the expert systems in training, and in the economy are given. Then we turn to the empirical study of the process of system's decomposition and to the algorithm for a comprehensive assessment of the effectiveness of the functioning of the innovation system based on the tensor estimation of system performance is also proposed in this chapter. We propose to use homomorphisms of the group $\boldsymbol{G_S}$ of factors defining the system S into the group $GL(n, \Delta)$ of linear homogeneous transformations of the vector space R^n as tensor estimates of the efficiency of the functioning of the system S. Also we can consider homomorphisms of the group of factors $\boldsymbol{G_S}$ that define a system S in the group $GL(n, \Delta)$ of linear homogeneous transformations of the vector space Δ^n over an arbitrary field Δ.

The following results were obtained as applications:

- economic systems. In an economics with the presence of the shadow sector, a system with full implementation of P-connections can not work autonomously if P does not implement any connections of the shadow sector.
- expert systems in learning. Testing with the full implementation of the links and the oral examination with the full implementation of the links give the same result. The levels of the impact of the knowledge system on the student have been singled out and tabulated. A more detailed consideration is given to one of the fragments of the knowledge system's decomposition, it makes possible to determine the exposure levels listed in the table. Expert systems, an algorithms for compiling a database of errors, an algorithm for compiling a knowledge base, a theorem on describing errors and a theorem on describing correct solutions have been used for this purpose. Together with the works [2–5], which deal with the issues of test quality and the practice of test assessment of knowledge in the Russian Federation, this makes it possible to determine the levels of knowledge of students with a sufficiently high degree of reliability.

6.1 Formalization of Innovation and Effectiveness Concepts

Let us present several definitions from [6–9] in order to provide consistency of notions and statements which are given below.

Definition 6.1 Let S be a system with integrity property P. Then system S is called an innovative system with deciding integrity property P or P-innovative system if system S can be off line in every super system S', such that S is P-pure in S', or, in other words, it contains it without a distortion of P-connections.

An innovative system S with integrity property P differs from a system with integrity property P higher (not lower than in super system containing it) implementation performance numeric innovative properties, or the fact that analogues of a system with property P does not exist. A system satisfying the property P will be called a P-system.

The main indicators of innovation are: newness, the degree or level of newness, consumer value, degree of implementation in practice, effectiveness, the presence of a single indicator of efficiency, the phenomenon of "flash" (a synergistic effect) characterizing the beginning of the autonomous work of the innovation system, the graph of all links of all innovations.

Basic Properties of Innovative Systems. Examples.

Now let's formulate and prove some theorems concerning innovative systems.

Let P be a predicate defining the integrity property of a system S. We shall consider P-pure embeddings in the class of all systems. From [7] we have the following main theorem.

Theorem 6.2 *An innovative system S with full realizations of P-connections not distorting P-connections of containing it super system is off line.*

Definition 6.3 Let $S = \{S_i | i \in I\}$—be a split of a main set of a system $\hat{A}$ with integrity property P, and $P(S_i)$ is true for every $i \in I$. The intersection $\cap_{i \in I} S_i = K$ is called the P—kernel of a system $\hat{A}$. As P is closed under intersections then $P(K)$ is true.

In practice $\{S_i | i \in I\}$ are different spheres of functioning operating system $\hat{A}$. For example, in the field of learning processes (learning processes) one can distinguish the following areas:

- the range of subject areas, in which, for example, are studied discipline with interdisciplinary linkages, and the like;
- the field of learning technologies, in which, for example, the following technology and generalized educational technologies are included: problematic instruction, concentrated training, developing training, modular training, differentiated instruction, active learning, 's training, and so on;
- the field of exit (field performance).

So the P-kernel of a system is essentially its infrastructure.[1]

In practice the spheres that are affected when the system $\hat{A}$ is functioning have as a rule non-empty intersection, and thus $\{S_i \cap A | i \in I\}$ is not a split of A. So one can consider $S' = \{S'_i | i \in I\}$, where $S'_i = S_i \backslash \cap_{i \in I} S_i$, $i \in I$, instead of $S = \{S_i | i \in I\}$. Then $\bigcap_{i \in I} S'_i = \emptyset$ and $\bigcup_{i \in I} S'_i = A$ and $S' = \{S'_i | i \in I\}$ is a split of A.

[1]Infrastructure is a set of interrelated service structures or objects, components and/or provide the basis for the functioning of the system

Theorem 6.4 *For a subsystem $\hat{B}$ of a system $\hat{A}$ to be an innovative one it is necessary that the main set B of the system $\hat{B}$ contains P-kernel of every its supersystem.*

As a consequence of Theorem 6.4 we have:

Theorem 6.5 *In an economy with a shadow sector, a system with full implementation of P-links cannot work autonomously if P does not implement any connections of the shadow sector.*

***P*-pseudo-innovative system**

Let's introduce the notion of *P*-pseudo-innovative system following the principle of duality.

Definition 6.6 Algebra $\bar{G} = \langle G | \{ f_\alpha^{n_\alpha} | \alpha \in \Gamma \} \rangle$ is called *P*-pure projective (*P*-pseudo-innovative system) if every diagram with exact *P*-pure string.

$$\begin{array}{ccccccc} & & & & \bar{G} = \langle G | \{ f_\alpha^{n_\alpha} | \alpha \in \Gamma \} \rangle & & \\ & & & \overset{\varphi}{\swarrow} & \downarrow \psi & & \\ 0 \to \bar{B} = \langle B | \{ f_\alpha^{n_\alpha} | \alpha \in \Gamma \} \rangle & \xrightarrow{\alpha} & \bar{A} = \langle A | \{ f_\alpha^{n_\alpha} | \alpha \in \Gamma \} \rangle & \xrightarrow{\pi} & \bar{C} = \langle C | \{ f_\alpha^{n_\alpha} | \alpha \in \Gamma \} \rangle & \to 0 & \end{array}$$

that is Im α is *P*-pure in $\bar{A}$ can be extended to commutative one that is $\psi\pi = \varphi$.

Theorem 6.7 *Direct (inductive) limit of P-pseudo-innovative systems is P-pseudo-innovative system.*

The proof of the theorem 6.7 pass like the corresponding ones in [1].

Hence, by analogy with [1], we obtain the theorem.

Theorem 6.8 *A system that is both P-innovative and P-pseudo-innovative, is P-degenerate, that is, it loses property P and cannot function.*

Corollary 6.9 *In order for the P-innovative system can function, it is necessary that it has no at least one property of the P-pseudo-innovation system.*

The meaning of this result runs as follows: a system without shortcomings does not work. The next question that arises in the course of studying the functioning of systems is the question of constructing an integrated assessment of the efficiency of the functioning of the smart system.

6.2 Algorithm for a Comprehensive Assessment of the Effectiveness of a Smart System

The task of constructing a numerical estimate of the effectiveness of the functioning of the system is extremely difficult from a mathematical point of view since its solution involves a quantitative assessment of the appearance of qualitative changes.

We shall construct a tensor estimate of the effectiveness of the functioning of the system as a homomorphism of a group of factors G_S, determining the system S into a group $GL(n, R)$ of linear homogeneous transformations of the vector space R^n.

In this connection we recall how a complete linear group is defined. We get the possibility of constructing various tensor estimates of the system's functioning efficiency by considering the homomorphisms of the group G_S, that define a system S into a group $GL(n, \Delta)$ of linear homogeneous transformations of the vector space Δ^n, where Δ is a finite field or a field different from R.

Let's recall the following definition.

Definition [10] A group $GL(n, \Delta)$ is called a full linear group of degree n over the ring Δ. It is isomorphic to the multiplicative group of the ring Δ_n, where Δ_n is a ring of all $n \times n$-matrices over the ring Δ.

Then $\Delta_n \cong EndM$ where M is a free right Δ-module with the finite basis. If $\sigma \in EndM$, then $\sigma(u_v) = \sum_{\mu=1}^{n} u_v \propto_{\mu v}, \propto_{\mu v} \in \Delta, v = 1, \ldots, n$. Let $A = \left\| \propto_{\mu v} \right\|$. A matrix A is called a matrix of endomorphism σ in the basis $u_1, \ldots, u_n$. Then the map $\varphi : EndM \rightarrow \Delta_n$ such that $\varphi : \sigma \mapsto A$, is an isomorphism.

An assessment of efficiency of functioning of the innovation system can be carried out according to the following algorithm.

6.2.1 *The Algorithm of a Complex Estimation of Efficiency of Functioning of the Innovation System*

Algorithm 6.10

(1) The construction of quantitative indicators describing the functioning of the innovation system. Identifying and monitoring the compliance of the functioning of the innovation system with its purpose.
(2) The construction of the tensor evaluation of the innovation system's functioning.
(3) The identification and monitoring of all the links of the innovation system with the external supersystem and its subsystems that arise during the functioning of the innovation system.
(4) The identification of internal and external attributive factors of the innovation system.
(5) The construction of internal qualitative indicators of the functioning of the innovation system in the form of a graph and a group of internal attributive features of the innovation system.
(6) The construction of external quality indicators of intrasystem connections of the innovation system in the form of a graph and a group of external attributive features of the innovation system.
(7) The determination of the number of possible synergetic effects of the innovation system.

(8) The construction a model an innovation system's functioning based on its internal and external attributive features.
(9) The verification of the possibility of increasing the accuracy of the model by introducing additional factors into the model.
(10) The correction the management actions to prevent the occurrence of undesirable synergistic effects.

The following innovation management model can be proposed based on the foregoing.

Model of innovation management
Innovation management can be implemented by using the following Algorithm 6.11.

Algorithm 6.11

1. For each innovation attribute define the external characteristics.
2. For each innovation attribute define the internal characteristics.
3. For each innovation-defined attribute characteristics.
4. The process of monitoring the implementation of each innovation: the definition of the beginning, the turning points in the development end of the innovation process described in Chap. 2; Algorithm of defining and regulating the properties of the system *S*. Other monitoring procedures described in [11].

The next question we shall consider is the quasi-stability of pseudo-innovation systems.

6.2.2 Quasi Sustainability of Pseudo-innovative Systems

Let us return to the consideration of *P*-quasi-stable systems and *P*-pseudo-innovation systems.

The numerical characteristics or the structural characteristics of the system *S* can vary within certain limits, however, the resulting changed system S' may satisfy to the property of integrity *P* of the system, that is $P(S) \Rightarrow P(S')$, and therefore it may to perform almost all the same functions as the system *S*, and, finally, to achieve the goal *Q*. Thus, it turns out that a change in the system *S*, possibly qualitative, possibly quantitative, within certain limits, does not violate its integrity. This property is called the quasi-stability of the system *S*.[2]

Definition 6.12 The system *S* is quasi-sustainable with respect to the property *P*, defining its integrity if there exists a system S' and a congruence $\equiv_P$ on S' such that the factor-model $S'/\equiv_P$ is isomorphic to *S*.

---

[2]According to Anthony Leonard Southern Corner, who introduced the notion of quasi-isomorphic torsion-free abelian groups.

Examples. Application to Expert Systems in E-Learning

We obtain from Theorems 6.2 and 6.4 as a consequence the following theorem.

Theorem 6.13 *Testing with full implementation of connections can work offline, that is to give an objective result.*

Under full implementation of connections we understand the following:

- availability of tests to verify all skills, knowledge and competence of the audited pupils,
- existence the control of self-dependence of fulfilling all of the tasks,
- the availability of an adequate system for verifying the tests for evaluating completed tasks.

It is necessary to have an error map, a map of the correct solutions for every test and executions of the algorithms proposed in [6].

In a similar way we obtain:

Theorem 6.14 *An oral exam with full implementation of connections can work autonomously, that is to give an objective result.*

So, we have the following theorem.

Theorem 6.15 *Testing with full implementation of connections and an oral exam with full implementation of connections give just the same result.*

Examples. Application to Economic Systems.

One of the most interesting issues that arises in connection with the study of the concept of a smart system is the question of the applicability of the "smart" ideology to such macrospheres in which artificial regulation, or regulation by a person or society, is currently possible at the microsphere level and micro processes, but their global implementation is impossible. A concrete examples of such a macrospheres, in our opinion, are such spheres as socio-economic sphere, a sphere of economic, a sphere of natural sciences. At micro-levels, of course, one can consider such socio-economic smart system subsystems as smart cities, smart training systems, smart healthcare, smart technologies, etc. Nevertheless, we can consider smart processes in such macrospheres.

In this connection, we shall give the following examples that characterize some aspects of smart processes in the economic.

We obtain from Theorems 6.2 and 6.4 as a consequence the following theorem.

Theorem 6.16 *The system of local tax authorities with the full implementation of links and, in particular, with full state control, can operate autonomously.*

On the other side we have Theorem 6.5 about shadow economics: In an economy with a shadow sector, a system with full implementation of P-links cannot work autonomously if P does not implement any connections of the shadow sector.

The following theorem is Sylow Theorem corollary about subgroups of a finite group.

Theorem 6.17 *If the closed associative system S is described with sufficient accuracy by a model consisting of m factors, then a model consisting of k factors,* *where $m|k$, can not be used to refine it, and there is no model specifying it from m factors, where m does not divide k, for example from $m+1$ factor.*

An example that empirically confirms the Theorem 6.17 is V. L. Makarov's model of six factors of linear regression which describes the Russian economics.

The next question that arises in this connection is the question of the independence of the variables appearing in the model. The answer to this question is given by the Theorem 6.18.

Theorem 6.18 *Let the closed associative system S be described with the help of the group of factors G_S. Then there are two independent variables x and y through which all the internal factors of the model G_S describing the system S are expressed.*

Proof Any group of rank n can be represented as a factor—group of a free group F_n of the rank n. Let's imbed F_n into a free group of a countable rank F_ω. Then let's imbed F_ω into $[F_2, F_2]$, where $[F_2,F_2]$ is a commutant of a free group of rank two. It is known that $[F_2, F_2]$ is a free group of a countable rank. So every element of a group G_S can be written as a combination of words of the form $w(a, b)H$, where a, b are independent variables, H is a normal subgroup of F_n. So each element of H can be expressed as a word from a and b.

6.3 Example. Decomposition of the Education System. Approaches to the Study of the Effectiveness of the Education System

Let's consider the effectiveness of the education system at first. Let's begin with example. We construct fragments of the process of decomposition of the education system. These fragments can be written in a Table 6.1.

Now let's decompose the system of education S into the following components:

1. Knowledge system S_1 (information subsystem)
2. Methodological and methodical complex S_2 of the knowledge system S (an adaptive subsystem)
3. The system of students S_3 (target subsystem, target audience).

The effectiveness of the knowledge system forms from the effectiveness of its subsystems S_1, S_2, S_3. The interaction of these subsystems may produce a synergistic effect. The efficiency function must take into account the possible synergistic effect. Let's describe the functioning and interaction, and the interaction of subsystems S_1, S_2, S_3 of the education system S.

Let's highlight the following basic steps for this.

Stage 1. S_2 effects onto S_1. Then an ordering of the knowledge system S_1 takes place. Let's designate it in the following way: $S_2(S_1)$ is a result of their interaction.

Stage 2. In turn, the result of this interaction $S_2(S_1)$ affects the target audience S_3.

Table 6.1 Table of estimation of effectiveness of the education system

		Performance criterion	Verbal description of the performance indicator	Measurer of the performance indicator	Notes designations
1.	Quality, including:	Compliance with the objective of the functioning of the system	Comprehensive assessment, including quantitative assessment indicators of the system functioning	Tensor evaluation of the effectiveness. It is also possible rating in the form of the utility function $t(S)$—utility function, $t(S) \to max$	$t(S)$—may be chosen also as system duration function, or $t(S)$—utility function, $t(S) \to max$
2.	The quality of management, or the quality of the education management system, including:	Accordance for purpose of the system of education management	Comprehensive assessment of the educational management system, including quantitative indicators of the performance evaluation of the system functioning	Tensor evaluation of the education management system, including quantitative indicators of the performance evaluation of the system functioning	To evaluate the quality of management, we construct a tensor estimate. We represent the tensor estimate as a homomorphism $G_{S_i} \to GL(n, R)$, where G_{S_i} is a group of factors defining the system S_i, $GL(n, R)$ is a group of linear homogeneous transformations of a vector space R^n, n—a number of quantitative indicators that assess the education management system
3.	The quality of the law framework, or the quality of the law subsystem providing the learning process	Correspondence to the purpose of functioning of the law subsystem of providing the learning process	Comprehensive assessment of the legal subsystem of the education system, including quantitative indicators of the evaluation of the functioning of the law subsystem of the education system	Tensor evaluation of law subsystems of the education system, including quantitative performance evaluation of the functioning of the law subsystems of the education system	To assess the quality of the law subsystem of the education system, we construct a tensor estimate. We represent the tensor estimate as a homomorphism$G_{S_i} \to GL(n, R)$, where G_{S_i} is a group of factors defining the system S_i, $GL(n, R)$ is a group of linear homogeneous transformations of a vector space

(continued)

Table 6.1 (continued)

		Performance criterion	Verbal description of the performance indicator	Measurer of the performance indicator	Notes designations
					R^n, n—a number of quantitative indicators that assess n—the number of quantitative indicators that assess the law subsystem of the education system
4.	Quality of financial support of the education system	Correspondence to the purpose of functioning of the financial subsystem providing the learning process	Comprehensive assessment of the financial subsystem of the education system, including quantitative indicators of the evaluation of the functioning of the financial subsystem of the education system	Tensor evaluation of financial subsystems of the education system, including quantitative performance evaluation of the functioning of the financial subsystems of the education system	To assess the quality of the law subsystem of the education system, we construct a tensor estimate. We represent the tensor estimate as a homomorphism $G_{S_i} \to GL(n,R)$, where G_{S_i} is a group of factors defining the system S_i, $GL(n,R)$ is a group of linear homogeneous transformations of a vector space R^n, n—a number of quantitative indicators that assess n—the number of quantitative indicators that assess the financial subsystem of the education system
5.	Qualifications of managers engaged in the management		Comprehensive assessment of the subsystem of managers responsible for the management of the education system, including quantitative indicators assessing the qualifications of managers.	Tensor assessment of the subsystem of managers who manage the education system, including quantitative indicators that assess the qualifications of managers	To assess the quality of the subsystem of managers responsible for the management of the education system, we construct the tensor estimate. We represent the tensor estimate as a homomorphism $G_{S_i} \to GL(n,R)$, where G_{S_i} is a group of factors defining the system S_i, $GL(n,R)$

(continued)

Table 6.1 (continued)

		Performance criterion	Verbal description of the performance indicator	Measurer of the performance indicator	Notes designations
					is a group of linear homogeneous transformations of a vector space R^n, n—the number of quantitative indicators that assess the quality of the subsystem of managers responsible for the management of the education system
6.	The quality of education, or the quality of the knowledge system, including:	Correspondence of the purpose of the knowledge subsystem	Comprehensive assessment of the knowledge system, including quantitative indicators that assess the knowledge system	Tensor evaluation of the knowledge system, including quantitative indicators that assess the knowledge system	To assess the quality of the knowledge subsystem of the education system, we construct a tensor estimate. We represent the tensor estimate as a homomorphism $G_{S_i} \to GL(n, R)$, where G_{S_i} is a group of factors defining the system S_i, $GL(n, R)$ is a group of linear homogeneous transformations of a vector space R^n, n—the number of quantitative indicators that assess the quality of the knowledge system

(continued)

Table 6.1 (continued)

		Performance criterion	Verbal description of the performance indicator	Measurer of the performance indicator	Notes designations
7.	Quality of methodological and (methodical) support	Correspondence to the level of knowledge of the contingent of students and compliance with training standards	The quality of tests and the assessment of the adequacy of control of students' knowledge in tests	Tensor evaluation of the system of methodological and (methodic) support	Quality of programs, courses, test materials, etc.
8.	Quality of teaching staff	Correspondence to the purpose of functioning of the pedagogical staff subsystem	Indicators assessing the qualifications of the teaching staff performing the educational process		
9.	Quality of feedback		The quality of tests and the assessment of the adequacy of control of students' knowledge by tests		

Table 6.2 Table of decomposition of the education system

Levels of exposure	Reactions of the target audience to the operation of the system $S_2(S_1)$		
1. Understanding $S_2(S_1)$	Understand	Don't understand	Not familiar with
2. Passive possession $S_2(S_1)$	Master the basic skills that characterize $S_2(S_1)$	Don't master the basic skills that characterize $S_2(S_1)$	Not familiar with basic skills that characterize $S_2(S_1)$
3. Active possession $S_2(S_1)$	Can solve tasks	Cannot solve tasks	Not familiar with tasks
4. Hold on the level of competence	Own a research approach: are able to solve tasks, to set tasks	Don't own a research approach: are not able to solve tasks, to set tasks	Not familiar with a research approach

Then the goal of training—the desired final state of the system is written as follows: $(S_2(S_1))(S_3)$.

Let's describe the reaction of the target audience S_3 to the action of the system $S_2(S_1)$. For this purpose we construct a table of reaction levels of the target audience for the action of the system $S_2(S_1)$. Let's select the following four levels in this table.

We shall continue the process of decomposition of the smart education system and consider in more details one of the fragments of the decomposition of the knowledge system, allowing to determine the levels of exposure listed in the Table 6.2. We shall use for this purpose expert systems and the algorithm for compiling a database of errors, the algorithm for compiling a knowledge base, the error description theorem, and the theorem on describing correct solutions given below. Note that works [2, 3, 5], examining the quality of tests and the practice of test assessment of knowledge in the Russian Federation, allow to determine the levels of knowledge of students, according to the Table 6.2.

6.4 Decomposition of the Knowledge System. The Representation of the System of Knowledge in the Form of an Algebraic System

Let us first consider the question of the representation of a system of knowledge. The following methods of presenting the knowledge system are known:

(1) Formal models of the representation of a system of knowledge;
(2) Non-formal models of the representation of a system of knowledge (semantic, relational).

Formal models are generally based on a fairly strict mathematical theory based on a logical output operation. Informal theories generally do not have a universality property and are created for specific queries. Output in informal systems is defined by the researcher. Each of these methods of knowledge presentation is consistent with its own way of describing knowledge. A formal model of knowledge representation is a formal theory.

$F = (A, V, W, R)$, that defines an axiomatic system, characterized by the presence of the alphabet A, a set of syntax rules V, the set of axioms W, defining the theory, the set R of inference rules. In formal models of representation of knowledge, it is difficult to take into account the dynamics of the simulated phenomenon or process. Therefore, they are used in areas that depend a little of external factors. Such informal models of knowledge representation as semantic networks are based on the properties of internal interpretability, structuredness and connectivity that make it possible to simplify the search for a solution to a task. Well-known expert systems have been created that use semantic networks as the language of representation of knowledge—PROSPECTOR, CASNET, TORUS.
Single out:

- logical models that allow to build solvable models of knowledge representation (with decidable elementary theory in the language of narrow predicate calculus),
- network models, that allow to perform all operations and make improvements into the structure of the model prior to its implementation,
- production models that allow to present knowledge in the form of sentences of the type: "IF the condition, THEN the action",
- frame models that allow the representation of all the properties of declarative and procedural knowledge.

A logical model is a formal system of the form $M = \langle T, S, A, B \rangle$, where T is the set of base elements of a different nature, and there is a way of determining whether an arbitrary element x belongs to the set T or not. The procedure for such a check of $P(T)$ in a finite number of steps must give a positive or negative answer to the question of whether x is an element of the set T. The set S is the set of syntactic rules with the help of which syntactically correct sets build from the elements of T. For example, from the words of a limited dictionary syntactically correct phrases are constructed, new constructions are assembled from the details of the child's designer using nuts and bolts. It is assumed that there exists a procedure $P(S)$ with the help of which in a finite number of steps one can get an answer to the question whether the set X is syntactically correct or not. A certain subset A of the axioms of the formal system is distinguished in a set of syntactically correct sets. It is assumed that there exists a procedure $P(A)$ with the help of which for any syntactically correct set it is possible to obtain an answer to the question whether it belongs to the set A or not. The set B is the set of inference rules. Applying to the set of inference rules to the elements of a set A, one can get new syntactically correct sets, to which one can again apply the rules from B. If there is a procedure $P(B)$ that can determine for a finite number of steps for any syntactically correct set, whether that syntactically correct set is deducible or not, then the corresponding formal system is called solvable, otherwise it is called unsolvable one. Thus, the formal system is a producer of the generation of new knowledge, forming a set of knowledge deduced in the given system. In fact, such actions present the output procedures described in Chap. 1. The network model displays the links between operations and the order of their execution. The operations are ordered in time. A wide range of tasks of optimization the planning and implementation of interrelated processes is solved with the help of network models. Tasks of this type arise in the realization of projects that include the implementation of a set of activities. The network model can be formally defined as a vector $H = \langle I, C_1, C_2, \ldots, C_n, \Gamma \rangle$ where I is the set of information units; $\{C_1, C_2, \ldots, C_n\}$ is the set of types of links between information units. The map Γ defines connections from a given set of connection types between the information units entering into I. Classification networks, functional networks and scenarios are distinguished by the types of connections used in the model. Structuring relationships are used in classifying networks. Such networks allow one to introduce different hierarchical relations between information units in knowledge bases. Functional networks are characterized by the presence of functional relationships. In scenarios, causal relations are used, as well as relations of types "means-result", "tool-action", etc. If the network model allows connections of different types, it is usually called a semantic network. Product models use elements of logical and network models. The idea of the rules of inference is borrowed from logical models, which here are called products, and the description of knowledge in the form of a semantic network is borrowed from network models. The semantic network is transformed by changing its fragments, building up the network and excluding unnecessary fragments from it as a result of applying the output rules to the fragments of the network description. Thus, procedural information is explicitly identified in production models. A knowledge-based conclusion appears in the production models instead of the logical inference which is characteristic of logical models. Frame model is a systematic psychological model of human memory and consciousness. A rigid structure of information units is fixed in frame models unlike other types of models. We propose to use as a model of knowledge representation algebraic systems that differ from formal models of representation of knowledge in that not every algebraic system can be described in the language of narrow predicate calculus with the help of an axiom system. Besides it algebraic systems used as knowledge models differ from informal knowledge models by the possibility to use the language of narrow predicate calculus of needed signature and by the possibility to use known facts about their structure [3, 6]. This allows to combine the advantages of formal and informal models of knowledge representation. It is possible to present a closed knowledge system as a form of a group of formalized correct actions to decide test tasks of knowledge system S. For this, the knowledge system should be represented by levels, that is, by the lattice of subgroups of the group which helps one formalizes the knowledge system S. Here S is a system of correct actions to solve test tasks. It can be done as follows. The expert writes out all the atomic statements $\{a_i | i \in I\}$ (statements in the signature Ω describing the knowledge system) that is statements that cannot be decomposed into statements of shorter length. Let's designate by a_i^{-1} the negation of a statement a_i. Then let's consider words in the alphabet $\langle \{a_i|i \in I\} \bigcup \{a_i^{-1}|i \in I\} \cup \emptyset$, where $\emptyset$ is an empty word, $\circ$ is an operation of sequentially ascription the words from left to right. Then an algebra $\langle \{a_i|i \in I\} \bigcup \{a_i^{-1}|i \in I\} \cup \emptyset | \square^{-1}, \circ, \emptyset \rangle$ is a group which represents the knowledge system. It is also possible to represent a closed knowledge system in the form of a Boolean algebra formalizing the statements of the knowledge system S. For each knowledge system, an expert writes out all atomic statements $\{a_i|i \in I\}$, that is statements that cannot be decomposed into statements of shorter length, in the signature Ω, describing the knowledge system in the narrow predicate calculus of the signature $\Omega = \langle 1, 0, \rceil \rangle$, where 1 designates true statement, 0 designates false statement. Then $\{\rceil a_{i_i}|i \in I\}$ is the set of all false atomic statements of the knowledge system S. If A and B are true statements then $A \bigwedge B$ is a true statement, $1 \wedge A \Leftrightarrow A \Leftrightarrow A \wedge 1$ and the algebraic system $\langle \{a_i|i \in I\}, \rceil, \Lambda, 1 \rangle$ is a Boolean algebra. Besides it one can consider it as a commutative semigroup with a unit.

6.5 Decomposition of the System. Analysis and Synthesis of the Knowledge Base

Definition 6.21 Let's consider the operation of decomposition of an algebraic system G which represents the knowledge system S, into the subsystems $\{G_i|i \in I\}$. Let's designate the operation of decomposition by the symbol $\coprod$, that is $G = \coprod_{i \in I} G_i$ is the designation of an algebraic system G onto subsystems $\{G_i|i \in I\}$. Let's designate the operation of system synthesis by the symbol $\prod$, that is $G = \prod_{i \in I} G_i$ is the synthesis of an algebraic system G from the subsystems $\{G_i|i \in I\}$. We shall introduce the order relation on the set I as follows: $i \precsim j \Leftrightarrow G_i \leq G_j$ for $i,j \in I$.

For every task an expert makes a complete set of errors with the following algorithm.

Algorithm 6.22 of errors description Algebraic formalization allows us to construct the following scheme of an Expert System for testing pupils in mathematics based on tools of Group Theory. Algorithm of compilation errors database runs as follows. Algorithm of compilation errors database is described by the language of the first order with the signature $\Omega = \langle *, \square^{-1}, e \rangle$ of Group Theory in the conjunction that a composition of mistakes is an associative binary operation. Let's use the following table of symbols.

Table of symbols: n is the number of tasks, $Q_1, Q_2, \ldots, Q_n$ are all tasks of database, search for tasks is carried out on the main menu, in the option section. Expert writes down all atomic errors each of which contains only one error action $\left\{m_j^i|i = 1, \ldots n, j = 1, \ldots, q_i\right\}$, where i is the number of a task, j is the number of an error for a given task. Composition that is sequential execution of actions of any errors, including atomic ones, is an error:

$m_j^i * m_r^s, i, s = 1, \ldots n, j, r = 1, \ldots, q_i$. If an atomic error m_j^i is made, then the reverse action is an error which will be designated by $(m_j^i)^{-1}$ or by m_j^{i-1}, and will be named a inverse error. Evidently $((m_j^i)^{-1})^{-1} = e$, where e is neutral error or neutral element. Pupil's answer on a test may be nominally true but it can contain neutral error, for example, $m_j^i * m_j^{i-1} = e$. Such answer could not be considered as the right decision. Thus the full system of errors of a given task can be described by a group $\boldsymbol{G} = \langle G, *, ^{-1}, e \rangle$ with the main set G, binary operation $*$, unary operation $^{-1}$ of taking inverse element and a neutral element e. We shall consider the operation of composition of errors to be an associative operation. Then $\boldsymbol{G}$ is a free group of the rank $k = n \sum_{i=1}^{n} q_i$ with generators $\left\{ m_j^i | i = 1, \ldots n, j = 1, \ldots, q_i \right\}$. The following theorem of errors' description is proved.

Theorem 6.23 of errors' description *The set of all errors is defined by no more than two combinations of words which are finite length's compositions of atomic errors.*

Proof Let's consider a free group of a finite rank $k = n \sum_{i=1}^{n} q_i$ with generators $\left\{ m_j^i | i = 1, \ldots n, j = 1, \ldots, q_i \right\}$. As every free group of a finite rank can be embedded into a free group of countable rank let's imbed $\boldsymbol{G}$ into a free group $\boldsymbol{F}_\infty$ of countable rank, and $\boldsymbol{F}_\infty$ imbed into $[\boldsymbol{G}, \boldsymbol{G}]$ which is a commutant of a group $\boldsymbol{G}$ and $[\boldsymbol{G}, \boldsymbol{G}]$ is a group of countable rank. Then let's imbed $[\boldsymbol{G}, \boldsymbol{G}]$ into $[\boldsymbol{F}_2, \boldsymbol{F}_2] \leq \boldsymbol{F}_2$ which is a commutant of a free group $\boldsymbol{F}_2$ of the rank two. It can be done because $[\boldsymbol{F}_2, \boldsymbol{F}_2]$ is a free group of a countable rank. So a group $\boldsymbol{G}$ can be embed into a free group $\boldsymbol{F}_2$ of rank two. Thus we have that there exist two words $w_1(m_i^i | i = 1, \ldots n, j = 1, \ldots, q_i)$ and $w_2(m_i^j | i = 1, \ldots n, j = 1, \ldots, q_i)$, that $\boldsymbol{G}$ can be embedded into

$$F_2 = < w_1(m_i^j | i = 1, \ldots n, j = 1, \ldots, q_i),\ w_2(m_i^j | i = 1, \ldots n, j = 1, \ldots, q_i) > .$$

Now consider search and analyses of mistakes and typification of errors and scaling. Experts can single out standard errors which can be written down as composition of atomic errors that is as words $\omega_\propto(x_1, \ldots, x_n), \propto \in \Gamma$ in an alphabet $\left\{ m_j^i | i = 1, \ldots n, j = 1, \ldots, q_i \right\}$, that is setting a group $\boldsymbol{G} = \langle G | \omega_\propto(x_1, \ldots, x_n) = e, \propto \in \Gamma \rangle$ with defining relationships—the group of errors of a task. Van Kampen's lemma allowed to see the group of errors of a task and to analyze it. Besides this one can build the subgroup lattice of this group giving the opportunity to represent graphically all errors in task number i solution.

Introduction of atomic errors allows to scale the area of errors by their length without taking into account neutral errors that is without species compositions relationships $a * a^{-1}$.

Algorithm for Compiling a Knowledge Base. Search and Analysis of Correct Task Solving Algorithms

Algorithm of compilation of knowledge base and search and analyses of true algorithms of decisions run as follows. For every task expert forms the full set of true actions by the following algorithm.

Table of symbols: n is the number of tasks, $Q_1, Q_2, \ldots, Q_n$ are all tasks of database, search for tasks is carried out on the main menu, in the option section. Experts can single out all atomic true actions $T = \left\{a^i_j | i = 1, \ldots, n, j = 1, \ldots, l_i\right\}$, where i is the number of a task, j is the number of a true action for a given task. A composition that is the consistent implementation of true actions, is a true action:

$$a^i_j * a^s_r, i, s = 1, \ldots n, j, r = 1, \ldots, l_i.$$

If an atomic true action a^i_j, is made then the reverse action is a true action is made then the reverse action is a true action which is designated by $(a^i_j)^{-1}$, or simply a^{i-1}_j. Evidently $((a^i_j)^{-1})^{-1} = e$.

The student's answer will formally be correct if his decision contains a neutral correct action, which is a composition of mutually inverse correct actions, i.e., for example $a^i_j * a^{i-1}_j = e$ is a neutral true action. So the full system of true actions to decide a task can be described as a closed algebra (closed system) $\boldsymbol{\Gamma} = \langle T, *, ^{-1}, e \rangle$ with a main set T, binary operation which is composition $*$, unary operation $^{-1}$ taking an inverse element, and a neutral element e.

We shall consider the composition of the correct actions as an associative operation. Then $\boldsymbol{\Gamma}$ is a free group of a finite rank $r = n \sum_{i=1}^{n} l_i$ with the set of generators $\left\{a^i_j | i = 1, \ldots n, j = 1 \ldots, l_i\right\}$.

Theorem 6.24 of true actions description *The set of all true actions is defined by no more than two combinations of words which are finite length's compositions of atomic true actions.*

Proof The proof of this theorem runs as the proof of the Theorem 6.2.

Let's consider a free group of a finite rank $k = n \sum_{i=1}^{n} q_i$ with generators $\left\{a^i_j | i = 1, \ldots n, j = 1, \ldots, q_i\right\}$. As every free group of a finite rank can be embedded into a free group of countable rank let's imbed $\boldsymbol{G}$ into a free group $\boldsymbol{F}_\infty$ of countable rank, and $\boldsymbol{F}_\infty$ imbed into $[\boldsymbol{G}, \boldsymbol{G}]$ which is a commutant of a group $\boldsymbol{G}$ and $[\boldsymbol{G}, \boldsymbol{G}]$ is a group of countable rank. Then let's imbed $[\boldsymbol{G}, \boldsymbol{G}]$ into $[\boldsymbol{F}_2, \boldsymbol{F}_2] \leq \boldsymbol{F}_2$ which is a commutant of a free group $\boldsymbol{F}_2$ of the rank two. It can be done because $[\boldsymbol{F}_2, \boldsymbol{F}_2]$ is a free group of a countable rank. So a group $\boldsymbol{G}$ can be embed into a free group $\boldsymbol{F}_2$ of rank two. Thus we have that there exist two words $w_1(a^i_i | i = 1, \ldots n, j = 1, \ldots, q_i)$ and $w_2(a^j_i | i = 1, \ldots n, j = 1, \ldots, q_i)$, that $\boldsymbol{G}$ can be embedded into $F_2 = < w_1(m^j_i | i = 1, \ldots n, j = 1, \ldots, q_i), w_2(a^j_i | i = 1, \ldots n, j = 1, \ldots, q_i) >$.

Now consider search and analyses of true actions and typification of true actions and their scaling.

Experts can single out standard true actions which can be written down as composition of atomic true actions that is as words $\{\mu_\beta(x_1,\ldots,x_n), \beta \in B\}$, in an alphabet $\left\{a_j^i | i = 1,\ldots n, j = 1,\ldots,l_i\right\}$, that is setting a group $\boldsymbol{G} = \langle G | \mu_\beta(x_1,\ldots,x_n) = e, \beta \in B\rangle$ with defining relationships—the group of true actions of a task. Van Kampen's lemma allows to see the group of true actions of a task and to analyze it. Besides this one can build the subgroup lattice of this group giving the opportunity to represent graphically all true actions in task number i solution.

Introduction of atomic true actions allows to scale the area of true actions by their length without taking into account neutral true action that is without species compositions relationships $a * a^{-1}$.

Now consider the analysis of solutions offered by the pupil.

Analysis of Pupils Solutions.

The pupil gives the full algorithm of solution of a task number i. Then the following procedure is performed:

(1) An expert writes down all atomic actions made by a pupil: $\{u_i^s | i = 1,\ldots,n; s = 1,\ldots,t_i\}$

(2) After that expert writes down all compositions of atomic actions of the pupil:

$\{v_\gamma(u_i^1,\ldots,u_i^{h_i}) | i = 1,\ldots,n, s = 1,\ldots,h_i\}$; after that an expert should consider the group $\boldsymbol{K}$, which is generated by all atomic actions fulfilled by a pupil: $\boldsymbol{K} = \langle\{v_\gamma(u_i^1,\ldots,u_i^{h_i}) | i = 1,\ldots,n, s = 1,\ldots,h_i\}\rangle$. So $\boldsymbol{K}$ is a group of a pupil's solution of a task. Van Kampen's lemma allowed to see the group of a decisions of a task and to analyze it. Besides it one can construct the lattice of all subgroups of this group giving an opportunity to portray all possible moves the pupils in solving the problem. This group may be countable, not necessarily finite. The decision given by the pupil is true if: $\boldsymbol{K} \subseteq \boldsymbol{\Gamma}; \boldsymbol{K} \cap \boldsymbol{M} = \langle \boldsymbol{e}\rangle$; where $\boldsymbol{M}$ is a group of false action and true answer is found out.

Transition to the Record of the Solution in the Language of the Narrow Predicate Calculus Language (NPC) from the Recording of the Solution in the Group Theory Language.

Writing down the true decisions and error decisions can be done in the first order language with the signature $\Omega = \langle *, \square^{-1}, e, \rangle$ when one examines the elementary theory of the group of true decisions of a task and the elementary theory of the group of errors of a task.

References

1. Fucks, L.: Infinite Abelian Groups, vol. 1, p. 335. Academic Press, New York and London (1970). And vol. 2, p. 416. Academic Press, New York and London (1973)
2. Serdyukov, V.I., Serdyukova, N.A.: New variants of information interaction among participants in the educational process. Uchenye zapiski IOO RAO **61**(1–2), 142–146 (2017). (in Russian)
3. Serdyukova, N.A., Serdyukov, V.I., Glukhova, L.V.: Algebraic approach to the system representation of knowledge in the intellectual automated learning and control system. Vector Sci. Togliatti State Uni. **33**(2), 328–335 (2015). (in Russian)
4. Serdyukov, V.I., Serdyukova, N.A.: Directions for improving computer knowledge control systems. Uchenye zapiski IOO RAO **61**(1–2), 147–150 (2017). (in Russian)
5. Serdyukov, V.I., Serdyukova, N.A.: Directions for the improvement of automated systems for monitoring the learning outcomes. Inf. Educ. Sci. **23**(3), 75–85 (2014). (in Russian)
6. Serdyukova, N.A., Serdyukov, V.I.: The new scheme of a formalization of an expert system in teaching. In: ICEE/ICIT 2014 Proceedings, paper 032, Riga (2014)
7. Serdyukova, N.A., Serdyukov, V.I., Slepov, V.A.: Formalization of knowledge systems on the basis of system approach, SEEL2015, In: Lakhmi, C., Jain, Hm, Uskov, V.L. (eds.) Smart Education and Smart e-Learning, Smart Innovation, Systems and Technologies, vol. 41, Springer, pp. 371–380 (2015)
8. Serdyukova, N.A., Serdyukov, V.I.: Modeling, simulations and optimization based on algebraic formalization of the system. In: 19th International Conference on Engineering Education July 20–24, Zagreb, Zadar (Croatia), New Technologies and Innovation in Education for Global Business, Proceedings, pp. 576–582, ICEE2015, Zagreb (2015)
9. Serdyukova, N.A., Serdyukov, V.I., Uskov A.V., Slepov V.A., Heinemann C.: Algebraic formalization of sustainability in smart university ranking system, In: Lakhmi, C., Jain, H., Uskov, V.L. (eds.) Smart Innovation, Systems and Technologies book series (SIST, vol. 75) International Conference on Smart Education and Smart E-Learning, pp. 459–474 (2017)
10. Suprunenko, D.A.: Matrix Groups, Science, Moscow (1972). (in Russian)
11. Glukhova, L.V., Serdyukova, N.A.: Multi-agent information system of risk management of innovation budgeting. Inf. Syst. Technol. Manage. Secur. **2**, 165–170 (2013). (in Russian)

Chapter 7
Algebraic Approach to the Risk Description. Linear Programming Models with Risk

Abstract One of the most important questions in the Systems Theory is the question of risks description. This chapter begins with an analysis of known existing approaches to risks description. In this connection, the main attention is paid to the quantitative definition of risk, which follows from the Kolmogorov-Chapman equation. Also, an approach to the classification of system risks from the position of algebraic formalization of the system is considered. The risk function r of the system is defined as a function dual to the probability measure in the framework of algebraic systems formalization. Probabilistic spaces with risk are examined. Some relationships between the Kolmogorov risk function $h(x)$ and the risk function $r(x)$ are found. The risks of changes in formalizations ("failure of formalization") of the system using the Kolmogorov-Chapman equation for exponential distribution are calculated. Also we present a model of linear programming with risk. In some cases a measure of systemic risk is proposed. Item content: a system S is a risk one or a P-risk one if it works in some cases autonomous, in the others not. Examples: P-innovative and P-effective systems are risk-free.

Keywords Risk function · Measure · Linear programming with risk

7.1 Introduction

The accumulation of new properties of the system is associated with bifurcations or with the appearance of a qualitatively different behavior of the system element when a quantitative change in its parameters takes place [1]. It is assumed that the probability of reliable prediction of new properties of the system is small at the time of bifurcation, where bifurcation is a kind of a system regeneration. Contradictions arise naturally in the process of system's development and they are the reason for the perfection development of systems. From the theory of systems it is known that it is impossible to speed up the development of the system by artificially introducing contradictions into it, since it is impossible to determine whether the system,

N. Serdyukova and V. Serdyukov, *Algebraic Formalization of Smart Systems*,
Smart Innovation, Systems and Technologies 91,
https://doi.org/10.1007/978-3-319-77051-2_7

as a result of their resolution, will bear the new qualities. In this regard, one of the most important questions in the Systems Theory is the question of risks description. In this chapter, an approach to the classification of system risks from the position of algebraic formalization of the system is considered. This approach made it possible to distinguish between regulated (internal) and unregulated (external) risks of a system. As it is known, the question about the existence of infinite systems is debatable one. However, in this framework it is shown that the set of unregulated risks of any infinite system has a power of continuum accurate up to the regulated risks. An algorithm for managing the internal regulated risks of the system is constructed for a system represented by a finite group of factors. The risk function r of the system is defined as a function dual to the probability measure in the framework of algebraic systems formalization. This allowed us to consider probabilistic spaces with risk (Definition 7.1). This chapter begins with an analysis of known existing approaches to risks description. In this connection, the main attention is paid to the quantitative definition of risk, which follows from the Kolmogorov-Chapman equation. Let us remind that the Kolmogorov-Chapman equation describes operations which occur to according to the scheme of Markov random processes. Some relationships between the Kolmogorov risk function $h(x)$ and the risk function $r(x)$ introduced in this chapter are found. Examples of the distribution functions $F(x)$ for which the risk function $h(x)$ is multiplicative one are considered. The risks of changes in formalizations ("failure of formalization") of the system using the Kolmogorov-Chapman equation for exponential distribution are calculated. The statistical definition of risk is considered. The chapter also presents a model of linear programming with risk, introduced by us in [2]. A linear programming model with a risk can be used in practice [2]. Examples of the use of algebraic formalization for describing systemic risk in the particular case when the system of factors determining the risk of a closed associative system is considered in conclusion. In the case when the factors determining the risk of the system form a complete group of events that are independent in aggregate at any time and have the same probability density satisfying a certain condition, a measure of systemic risk is proposed.

7.2 Known Approaches to the Mathematical Determination of Risk. The Kolmogorov Risk Function

If the essential (qualitative) definitions of risk are still not fully developed, then the computational aspects of risk have been studied quite well nowadays. The most well-known approach to the quantitative definition of risk is the definition of risk, which follows from the Kolmogorov-Chapman equation. The Kolmogorov-Chapman equation appears in the modeling of operations by the scheme of Markov random processes [3]. A random process proceeds in the system S, if the state of the system changes in time in an accidental, unpredictable manner. A random process occurring in a system S is called a Markov process or a process without aftereffect if it has the following property:

For each time t_0 the probability of any state of the system S in the future (for $t > t_0$) depends only on the state of the system S at the moment and does not depend on the history of the process i.e. on that when and how the system reached its present state. Markov processes are classified into random processes with discrete time and random processes with continuous time, depending on at what moments of time—in advance or known—there can be system transitions from one state to another [3].

The Kolmogorov-Chapman equation [4] for continuous linear operators $P(t), t > 0$, expresses a semigroup property in a topological vector space and is written in the matrix form as:

$$P(t+s) = P(t)P(s) \tag{7.1}$$

Most often the term "Kolmogorov-Chapman equation" is used in the theory of homogeneous Markov random processes, where $P(t), t \geq 0$ is the operator that transforms the probability distribution at the initial instant of time into a probability distribution at time t. ($P(0) = 1$).

Formally differentiating Eq. (7.1) with respect to s for $s = 0$, from equation

$$\frac{P(t+s) - P(t)}{s} = P(t)\frac{P(s) - I}{s}$$

we obtain the direct Kolmogorov equation:

$$\frac{\mathrm{d}P(t)}{\mathrm{d}t} = P(t)\Lambda, \quad \text{where } \Lambda = \lim_{s \to 0} \frac{P(s) - I}{s} \tag{7.2}$$

where I is the identity operator.

Formally differentiating Eq. (7.1) with respect to t for $t = 0$ we obtain the inverse Kolmogorov equation:

$$\frac{\mathrm{d}P(t)}{\mathrm{d}t} = \Lambda P(t), \text{ wherein } P(0) = I \tag{7.3}$$

Each of Eqs. (7.2) and (7.3) has a unique solution $P(t) = e^{\Lambda t}$, where

$$e^{\Lambda t} = I + \sum_{k=1}^{\infty} t^k \Lambda^k / k!$$

For infinite-dimensional spaces, the operator Λ is not necessarily continuous, and it can be defined not everywhere, for example, to be a differential operator in the space of distributions.

Equation (7.1) is the matrix form of the system of equations

$$\begin{aligned} & p_{ij}(t+s) = \sum_k p_{ik}(s) \cdot p_{kj}(t) \\ & i,j \in C \end{aligned} \tag{7.4}$$

The equation $p_{ij}(t+s) = \sum_k p_{ik}(s) \cdot p_{kj}(t)$ expresses the fact that the Markov system passing from the state i to the state j in time $t+s$, first in time s from the state i passes into some intermediate state k, and then in a time t from the state k goes to the state j, and the probability of the second transition does not depend on how the state k was achieved. In fact in Eq. (7.1) $P(t)$ is the transition probability matrix, $\Lambda = \|\lambda_{ij}\|$ is an infinitesimal matrix,

$$\Lambda = \lim_{s \to 0} \frac{P(s) - I}{s},$$

$P(s)$ is matrix of transition probabilities, λ_{ij} is the intensity (density) of the transition of the Markov chain from state i into state j.

From the Kolmogorov-Chapman equation, the risk function $h(x)$ is defined as follows:

$$h(x) = \frac{f(x)}{1 - F(x)}$$

where $f(x)$ is a function of probability density, $F(x)$ is a probability distribution function.

For absolutely continuous random variables we have:

$$h(x) = \frac{\mathrm{d}F(x)}{\mathrm{d}x} \cdot \frac{1}{1 - F(x)}$$

Then

$$\begin{gathered} \int h(x)\mathrm{d}x = \int \tfrac{\mathrm{d}F(x)}{\mathrm{d}x} \cdot \tfrac{1}{1-F(x)} \mathrm{d}x = -\ln(1 - F(x)) + c \\ 0 \le F(x) < 1 \end{gathered}$$

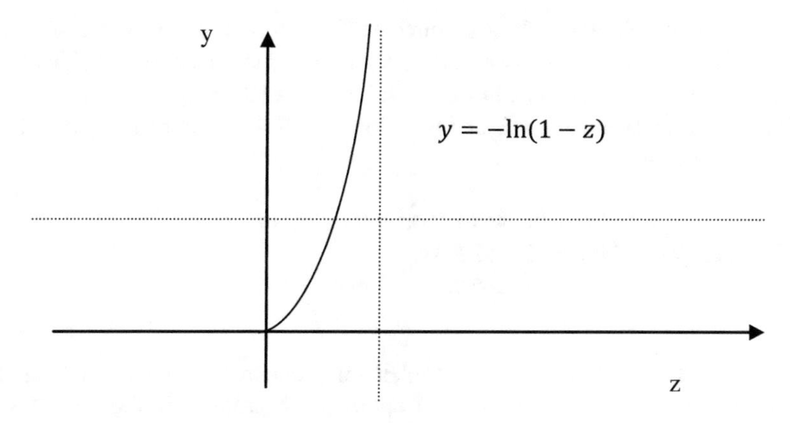

Fig. 7.1 Function graph $y = -ln(1 - z)$ on $[0; 1)$

Statistical risks It is also used such a quantitative risk assessment as a statistical risk. In the statistical theory of decision-making, the risk function for estimating $\delta(x)$ for the parameter θ, calculated for some observable x, is defined as the mathematical expectation of the loss function L (Fig. 7.1):

$$R(\theta) = \int L(\theta, \delta(x)) \times f(x|\theta)\mathrm{d}x, \tag{7.5}$$

where $\delta(x)$ is an assessment, θ is a valuation parameter. In applied fields such as economics and finance in the risk modeling use game theory, linear programming, linear programming with risks [2]. In [2] we introduced the concept of a linear programming model with risk, which then was applied to the optimization problem for VPN networks. Let us dwell on this model in more detail.

7.3 The Presentation of the General Model of Multi-criteria Optimization Problem in the Form of Linear Programming Task with Risks

Let's present the general model of multi-criteria optimization problem in the form of a linear programming taking the risks into an account. Let us write the multi optimization problem in a general form.

It runs as follows.

Multi-criteria optimization problem:

$$\begin{cases} \bigwedge W_j(x_1, x_2, \ldots, x_n) \to \max(\min), & j = 1, \ldots, m \\ \bigwedge P_i(x_1, x_2, \ldots, x_n), & i = 1, \ldots, k \end{cases} \tag{7.6}$$

Let's explain this entry. It is required to find the extremum of the objective functions $W_j(x_1, x_2, \ldots, x_n), j = 1, \ldots, m$, which are local as a rule and with the restrictions $\bigwedge P_i(x_1, x_2, \ldots, x_n)$, $i = 1, \ldots, k$ on the variables $x_1, x_2, \ldots, x_n$.

Step 2. Replacement the constrains in the optimization problem by risk functions. For each condition

$P_i(x_1, x_2, \ldots, x_n)$, $i = 1, \ldots, k$ we construct risk function r_i, $i = 1, \ldots, k$, and consider acceptable boundaries of risks:

$$a_i \leq r_i \leq b_i, \quad i = 1, \ldots, k$$

Step 3. Recorded the original optimization problem in the form, taking into account the risk: Recorded the original optimization problem in the form, taking into account the risk:

$$\begin{cases} \bigwedge W_j(x_1, x_2, \ldots, x_n) \to \max(\min), & j = 1, \ldots, m \\ a_i \leq r_i \leq b_i & i = 1, \ldots, k \end{cases} \tag{7.7}$$

Let's write the following simplification for (7.6) and therefore for (7.7). We shall designate by $R_j : R^n \to R$, risk's function for $W_j(x_1, x_2, \ldots, x_n)$, $j = 1, \ldots, k$. After that one must choose reasonable boundaries for every function $R_j, j = 1, \ldots, k$. Let $A_j \leq R_j \leq B_j$, $j = 1, \ldots, k$.

This can be done by the method of expert evaluations, or as in [5], etc. So we have for (7.7) and therefore for (7.6) if $j = 1,\ 2,\ 3\ or\ j = 1,\ 2$:

$$\begin{cases} \bigwedge A_j \leq R_j \leq B_j, & j = 1,\ 2,\ 3\ or\ j = 1,\ 2 \\ \bigwedge a_i \leq r_i \leq b_i, & i = 1, \ldots, k \end{cases} \tag{7.8}$$

In a more general form, when the optimization problem contains m objective functions, (7.7) can be put as follows:

$$\begin{cases} \bigwedge A_j \leq R_j \leq B_j, & j = 1, \ldots, m \\ \bigwedge a_i \leq r_i \leq b_i, & i = 1, \ldots, k \end{cases} \tag{7.9}$$

The physical meaning of (7.9) is described with the help of following task. One should find the mutual part of two n-dimensional polygons $\bigwedge A_j \leq R_j \leq B_j, j = 1, \ldots, m$ and

$\bigwedge a_i \leq r_i \leq b_i, i = 1, \ldots, k$, and choose any point in this mutual part. Let's illustrate (7.9) (Picture 7.1):

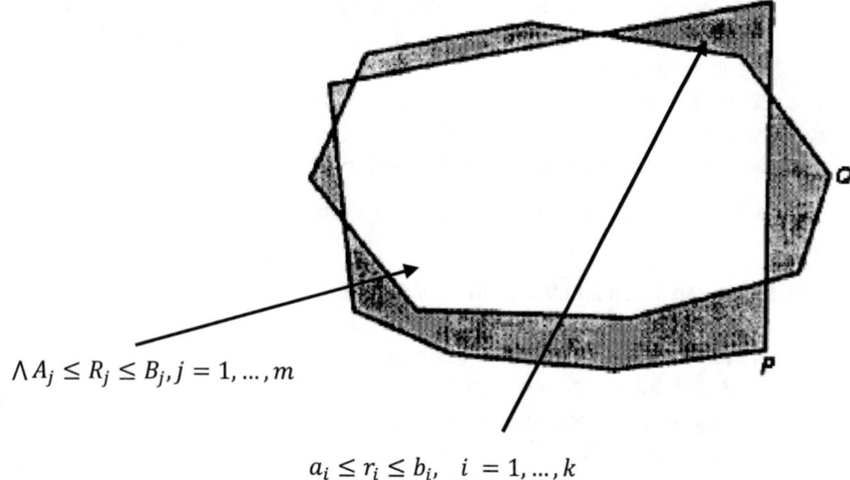

Picture 7.1 Optimization problem for dimension 2 in the form of linear programming task with risks

7.4 System Approach to Risk

Eelier using the language of the theory of algebraic systems we have formalized the term "system" as an aggregate of elements structured by connections into subsystems of different levels for reaching goals of functioning. Also the property of integrity of a system was formalized in the language of the theory of algebraic systems. In this regard it is necessary to be able to recognize the moments when the system changes its properties. Let's begin with examples.

Bifurcation[1] system risk is an unplanned or a planned with little probability transition from one status quality to another. Let's single out features of beginnings of a risk:

- a break of intercommunications of a system. A break of lows of a functioning of elements of a system is down.
- a break in *h* intercommunication high levels of hierarchy—in the centers of treatment of generalized information and working out decisions.
- a break of information's channels between units of a system or their overwork because it leads to difficulties in exchange of information between units of a system.
- a break of reliability of information used by a system.

[1]Bifurcation is the acquisition of a new quality in the movements of a dynamic system under small change of its parameters. Fundamentals of the theory of bifurcations of the Poincare lay in the early twentieth century.

- a break of quality of working of executive units and decreasing of resources lead to weakening of management and as a consequence to distortion of all dynamic process in the system.
- a management over a system is possible on the assumption of trustworthy forecast of a developing of the situation.
- a high level of external impacts over a system.

So, we have classification of risk by main moments of an appearance in the system:

Risk of a break of intercommunications of a system.
Risk of a break of lows of a functioning of elements of a system is down.
Risk of a break in *h* intercommunication high levels of hierarchy.
Risk of a break of information's channels.
Risk of a overwork of information's channels.
Risk of a break of reliability of information the system used.
Risk of a break of a quality of executive units.
Risk of a decreasing of resources the system used.
Risk of a distortion of dynamic process in a system.
Risk of exceeding an admissible level of external impacts over a system.
Risk of a distortion a forecast of a behavior of a system, etc.

A lot of examples shows one the necessity of introduction of universal risk function.

To ensure generality we shall use an axiomatic approach to define the function of risk.

7.5 Mathematical Model of Risk

Now the common definition of risk is absent. Let's start building the risk function. We shall define a function of risk as a function dual to probability measure.

Definition 7.1 Let an ordered triple $\langle \Omega, \Sigma, p \rangle$ be a probability space. A function $r : \Sigma \to R$, where R is a set of real numbers, is called a risk function if the following conditions are realized:

(1) for any countable set $\{A_i \in \Sigma | i \in N\}$

the following equation is true

$$r\left(\bigcap_{i \in N} A_i\right) = \prod_{i \in N} r(A_i) \text{ if } \bigcup_{i \in N} A_i = \Omega \tag{7.10}$$

(2) risk of impossible event is equaled to 0: $r(\varnothing) = 0$

An ordered quad $\langle \Omega, \Sigma, p, r \rangle$ is a probability space with risk.

Function of measure and particular a probability measure and risk function are differs by their properties, so risk cannot be defined as a measure.

Now, let's define risk of a random variable.

Definition 7.2 Let ξ be a variable in the probability space $\langle \Omega, \Sigma, p, r \rangle$ with a risk function, i.e. $\xi : \Omega \to R$ satisfies the following condition: for any $x \in R$ the set $\{\omega | \xi(\omega) < x\} \in \Sigma$ that is an event. Risk of a variable ξ is a function $r_\xi : R \to R$ such that $r_\xi(x) = r(\{\, \omega | \xi(\omega) < x\}$ for every $x \in R$.

Let ξ be a variable in the probability space $\langle \Omega, \Sigma, p, r \rangle$ with a risk function, and $A \in \Sigma$ is an event. A function $\chi_A : \Omega \to \{0, 1\}$ where

$$\chi_A(\omega) = \begin{cases} 1, \omega \in A \\ 0, \omega \notin A \end{cases} \tag{7.11}$$

is an indicator of A. Then

$$\overline{\chi_A(A)} = \int \chi_A dp = p(A) \tag{7.12}$$

A dispersion of an event A is

$$\sigma_A^2 = \overline{(\chi_A(A) - pA)^2} = \overline{\chi_A^2 - 2\chi_A p(A) + p^2(A)} = \overline{\chi_A^2 - p^2(A)} \tag{7.13}$$

Let's consider examples to show that some known quantity descriptions of risk are satisfied to this scheme. These are standard deviation, generalized variance and so on.

To use risk function not only for a system of pair wise independent events but in the common case for every probability space $\langle \Omega, \Sigma, p, r \rangle$ we construct a probability space $\langle \Omega^*, \Sigma^*, p^* \rangle$ containing the first one and such that there are natural embeddings $\Omega \to \Omega^*$, $\varepsilon : \Sigma \to \Sigma^*$ and $p^* \lceil \varepsilon(A) = p$ in which every two events different from certain event are independent. To construct such space we use the notion of inverse limit of inverse spectrum of factor algebras of σ-algebras by it's ideals, more precisely the inverse limit of inverse spectrum of probability spaces.

In the next section the theorem on the description of the passwords in the hierarchy of alphabets is proved. It is an important part for us because it resonates with the description of scales and together they show the importance of the free group of the rank two as a carrier of all useful information.

It follows from Definition 7.1 that the theory of infinite products can be used for the quantitative determination of risks.

7.6 The Use of the Theory of Infinite Products to Quantify Risks

Let's recall the basic definitions associated with infinite products [6].

Definition 7.3 Let $r_1, r_2, \ldots, r_n, \ldots$ be some sequence of real numbers. Symbol

$$r_1 \cdot r_2 \cdot \ldots \cdot r_n \cdot \ldots = \prod_{n=1}^{\infty} r_n \tag{7.14}$$

is called an infinite product. A sequence of partial products $\{P_n\}$, where $P_1 = r_1, P_2 = r_1 \cdot r_2, \ldots, \quad P_n = r_1 \cdot r_2 \cdot \ldots \cdot r_n, \ldots$ is called a sequence of products. The limit of a partial product (finite or infinite) $P = \lim_{n \to \infty} P_n$ is called the product value of $\prod_{n=1}^{\infty} r_n$. The value of the product is denoted as follows:

$$P = r_1 \cdot r_2 \cdot \ldots \cdot r_n \cdot \ldots = \prod_{n=1}^{\infty} r_n$$

If an infinite product has a finite value different from zero, then the product is called convergent, otherwise the product is said to be divergent.

$$\pi_m = r_{m+1} \cdot r_{m+2} \cdot \ldots \cdot r_{m+k} \cdot \ldots = \prod_{n=m+1}^{\infty} r_n \tag{7.15}$$

is called a residual product.

The following statements are well known:

(1) if the product (7.14) converges, then the product (7.15) converges for any m,
(2) if the infinite product (7.14) converges, then

$$\lim_{m \to \infty} \pi_m = 1$$

(3) if the infinite product (7.14) converges, then

$$\lim_{n \to \infty} r_n = 1$$

In the case of a convergent product, the factors r_n, beginning with some number, by virtue of (3), will be positive. Therefore, we can assume without loss of generality that all $r_n > 0$ for a convergent product.

(4) An infinite product (7.14) converges if and only if the series

$$\sum_{n=1}^{\infty} ln r_n \tag{7.16}$$

is a converged one.

If this condition is satisfied, and L is the sum of the series (7.16), and L_n is a partial sum of the series (7.16), then

$$P = e^L, L_n = \ln P_n, P_n = e^{L_n}.$$

If one sets $r_n = 1 + a_n$, then the infinite product (7.14) can be written in the form

$$\prod_{n=1}^{\infty}(1 + a_n) \tag{7.17}$$

and the series (7.16) can be written in the form $\sum_{n=1}^{\infty} \ln(1 + a_n)$.

Then we have:

(5) If the inequalities $a_n > 0$ or $a_n < 0$ are satisfied for sufficiently large n then the product (7.17) convergences if and only if the series $\sum_{n=1}^{\infty} a_n$ is a convergence one.

7.7 The Connection Between the Kolmogorov Risk Function *h*(*x*) and the Risk Function *r*

Let us now return to how $h(x)$ and the risk function r introduced in Definition 7.1. are related with each other. To get an answer to this question it is necessary to check in which cases the risk function $h(x)$ is a multiplicative one. Thus we should answer the following question: for what functions $F(x)$ the function

$$h(x) = \frac{f(x)}{1 - F(x)} = \frac{\mathrm{d}F}{\mathrm{d}t} \cdot \frac{1}{1 - F(x)} = \frac{\mathrm{d}}{\mathrm{d}t}(-\ln(1 - F(x))$$

is a multiplicative one. Indeed from condition (1) of Definition 7.1 follows that the risk function r is multiplicative. Recall that a function f defined on a certain set X is said to be multiplicative if for any $x, y \in X$ the equality $f(x \cdot y) = f(x) \cdot f(y)$ holds.

In order to avoid inconsistency in the definitions, let's note that in the Theory of Numbers, the multiplicative function is the arithmetic function $f(m)$ which is defined on the set of natural numbers and is taking values in the set of complex numbers, such that for any mutually prime numbers m and n the equality $f(m \cdot n) =$

$f(m) \cdot f(n)$ is true and $f(1) = 1$. Therefore, we shall look for examples of such functions $F(x)$ for which the risk function $h(x)$ is multiplicative.

7.7.1 The Simplest Examples of Probability Distributions with a Multiplicative Risk Function

Well-known examples of multiplicative functions are:

- the number of divisors $\tau(m)$ of the number m,
- the Mobius function $\mu(m)$,
- Euler function $\varphi(m)$, which is defined as the number of numbers from the series 0, 1, 2, …, $a - 1$, relatively prime to a,
- a power function $f(x) = x^{\propto}$.

1. **Power distribution.** In 1897, V. Pareto showed that events with a low effect occur frequently, and that events with an extremely large effect occur very rarely. In particular, he showed that 20% of the country's population controls 80% of its wealth. Power distribution violates this rule. In accordance with the power distribution of the event with a very large effect can occur quite often. From the theory of extremal values it is known that in complex nonlinear systems near the points of crises there is a power distribution known even under the names "distribution with a heavy tail", "Zipf-Pareto distribution" in linguistics, "law 20/80", "$1/f$-distribution", "fractal distribution". The power distribution is given by the following formula:

$$F(X) = \begin{cases} 1 - x^{-\propto}; & x \geq 1 \\ 0; & x < 1 \end{cases} \quad \propto > 0$$

The Pareto distribution is found not only in the economy. In linguistics, the Pareto distribution is known as Zipf's law, showing the relationship of the absolute frequency of words (how many times each word was found) in a sufficiently long text to the rank (sequence number when sorting the words according to the absolute frequency). Power law remains irrespective of, whether the words are to initial the form or taken from the text as it is. A similar curve occurs for the popularity of the names. The distribution of the size of settlements is power distribution. The distribution of file size in Internet traffic over the TCP protocol is a power distribution. In [7] the following illustration for the power distribution is given (Picture 7.2):

Let's compute $h(x)$. We have:

$$h(x) = \frac{f(x)}{1 - F(x)} = \frac{F'(x)}{1 - F(x)} = \begin{cases} \propto \frac{x^{-\propto-1}}{x^{-\propto}} = \propto x^{-1}, & x \geq 1 \\ 0, & x < 1 \end{cases}$$

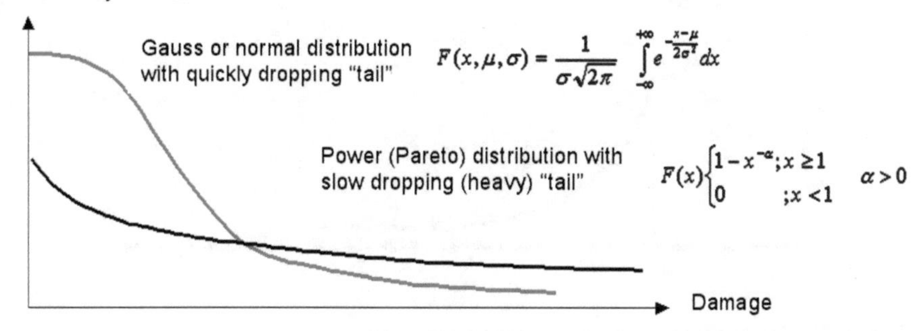

Picture 7.2 Power distribution with slow dropping "tail" and Gauss distribution with quickly dropping "tail"

It follows from this that in the case of a power distribution the function $h(x)$ is multiplicative.

2. **Exponential distribution**. Exponential distribution is widely used in reliability theory. Its characteristic property is the absence of aftereffect, i.e. in the case of the exponential law, the probability of failure-free operation of the device "in the past" does not affect the probability of its failure-free operation "in the near future." An exponential distribution also plays an important role in the theory of Markov random processes, queuing theory. The discrete analogue of the exponential distribution is the geometric distribution that is the distribution of a discrete random variable equal to the number of trials of a random experiment until the first "success" is observed.

The exponential distribution is given by the density

$$f(x) = \begin{cases} 0, & x<0 \\ \lambda e^{-\lambda x}, & x \geq 0 \end{cases}$$

The distribution function has the following form:

$$F(x) = \int_{-\infty}^{x} f(t)\mathrm{d}t = \int_{-\infty}^{0} 0\mathrm{d}t + \int_{-\infty}^{x} e^{-\lambda t}\mathrm{d}t = 1 - e^{-\lambda x} = \begin{cases} 0, & x<0 \\ 1 - e^{-\lambda x}, & x \geq 0 \end{cases}$$

Let's calculate $h(x)$. We have:

$$h(x) = \frac{f(x)}{1 - F(x)} = \frac{F'(x)}{1 - F(x)} = \frac{\lambda e^{-\lambda x}}{1 - (1 - e^{-\lambda x})} = \frac{\lambda e^{-\lambda x}}{e^{-\lambda x}} = \lambda$$

For $\lambda = 1$, the function $h(x)$ is multiplicative (Pictures 7.3 and 7.4).

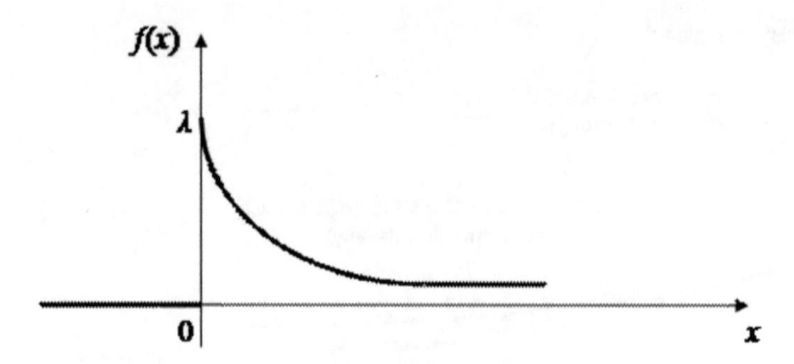

Picture 7.3 Density function for exponential distribution

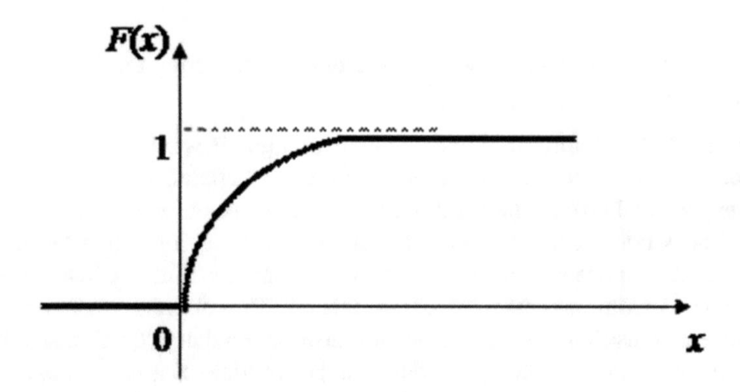

Picture 7.4 Distibution function for exponential distribution

7.8 Regulated Risks. Semigroup of Systemic Risks. Description of the System's Risk Semigroup

Definition 7.4 Let's call the risk $r:\Sigma \to R$ adjustable if $\lim_{i\to\infty} r_i = 1, r_i = r(A_i)$ for any countable set $\{A_i \in \Sigma | i \in N\}$.

If this condition is not fulfilled, then we call the risk $r:\Sigma \to R$ catastrophic or unregulated.

Definition 7.5 The internal risk of the system or systemic risk is defined as the risk determined by the internal attributes of the system, Chap. 4, Definition 4.14.

Definition 7.6 External risk of the system is defined as the risk determined by the external attributes of the system, Chap. 4, the Definition 4.15.

7.8.1 Simulation of the Moment of the Crisis of the System with the Help of the Kolmogorov-Chapman Equation

Let's consider the simulation of the moment of the crisis of the system with the aid of the simplest flow and the Kolmogorov-Chapman equation.

To solve problems of synthesizing optimal methods for receiving and processing information in laser information systems, discrete distribution laws are used as probabilistic models. At present, as the literature analysis shows, only 15 types of discrete distribution laws are known, while more than a thousand continuous laws are known [8]. As it often occurs, in order to obtain simpler expressions and simplify the computations and block diagrams of the processing devices, one must make the transition from discrete to continuous distribution laws.

7.8.2 Risks of Formalization Changes for the Exponential Distribution

We calculate the risks of changing formalizations using the Kolmogorov-Chapman equation. The exponential distribution is the only absolutely continuous distribution for which the no-after-effect property is satisfied. In this sense, the exponential distribution is a continuous analog of the discrete geometric distribution.

The probability distribution series of the geometric distribution has the following form:

The value of a random variable X	1	2	…	m	…
Probability p	p	qp	…	$q^{m-1}p$	…

Here X is the number of the first success in the test sequence. Let success means a change of formalizations, which we shall consider as a consequence of the crisis that has arisen in the system, since the description of the system with the help of the "old" formalization does not satisfy the already changed conditions. Let the random event consist in the fact that the formalization $\boldsymbol{A}_i$ is replaced by the formalization A_{i+1}. Then $h(x) = \lambda$, whereλ is the failure rate. The main issue here is how in practice make it possible to establish the failure of one or another formalization in describing the system.

Algebraic formalization	A_1	A_2	…	A_k	…
Probability	p_1	p_2	…	p_k	…

7.8.3 Algebraic Approach to the Description of Risks. Internal and External Systems Risks. Systemic Risk or System Risk

Definition 7.7 System S is called risky, if in some cases it works autonomously, but in others it does not.

The concept of risk can be defined more precisely if we associate the risk with the P-property of the system.

Definition 7.8 System S is called P-risky, if in some cases it works autonomously, but in others it doesn't.

Definition 7.9 Let x be an arbitrary real number, and : $\Sigma \to R$ be a risk function for the σ-algebra $\langle \Sigma, \cup, \cap, ', \Omega, p \rangle$ with a probability measure p defined on it. For each real number x we set

$$\Sigma_x = \{A | A \in \Sigma \backslash \{\emptyset\}, r(A) \leq x\}$$

$$\Sigma^1 = \{A | A \in \Sigma, r(A) = 1\}$$

Theorem 7.10 *For any real number x, the* algebra $\langle \Sigma_x, \cap \rangle$ *is a commutative semigroup with identity.*

Corollary 7.11 *The algebra* $\boldsymbol{\Sigma}_1 = \langle \{A | A \in \Sigma \backslash \{\emptyset\}, r(A) \leq 1\}, \cap \rangle$ *is a commutative semigroup with identity.*

The algebra $\boldsymbol{\Sigma}^1 = \{A | A \in \Sigma, r(A) = 1\}, \cap$ *is a commutative semigroup.*

The algebra $\boldsymbol{\Sigma}^1$ is called the semigroup of regulated risks of the system S.

Theorem 7.12 *The risk function* $\langle r : \Sigma \backslash \{\emptyset\}, \cap \to R \backslash \{0\}.\cdot \rangle$ *is a homomorphism of the commutative semigroup* $\langle \Sigma \backslash \{\emptyset\}, \cap \rangle$ *of the system S risks into the commutative multiplicative semigroup of non-zero real numbers* $\langle R \backslash \{0\}, \cdot \rangle$. *The kernel* $Ker\, r$ *of the homomorphism* r *is the semigroup of regulated risks* $\langle \boldsymbol{\Sigma}^1 = \{A | A \in \Sigma, r(A) = 1\}, \cap \rangle$.

From Theorem 7.12, we obtain by the theorem about homomorphisms

$$\langle \Sigma \backslash \{\emptyset\}, \cap \rangle / Ker\, r \cong Imr \leq \langle R \backslash \{0\}, \cdot \rangle$$

or

$$\langle \Sigma \backslash \{\emptyset\}, \cap \rangle / \boldsymbol{\Sigma}^1 \cong Imr \leq \langle R \backslash \{0\}, \cdot \rangle$$

that is, the set of unregulated risks of any system has a power no more than a continuum up to regulated risks.

7.8.4 *Algorithm for Regulating the Internal Risks of the System*

Let's consider firstly the examples of risks functions on a finite σ-algebras.

Example

1. First of all let's consider an extreme case. Let the system S is described by a finite group G_S, which is consist of two factors. Then $G_S \cong Z_2$ is a finite cyclic group of the order 2 with the following Cayley table

	0	1
0	0	1
1	1	0

$\Sigma = \langle \{\emptyset, \{0\}, \{1\}, \{0,1\}\}, \cup, \cap, \backslash, \prime \rangle$ is a σ-algebra for system S risks' description

Let

$$r(\emptyset) = 0,$$
$$r(\{0\}) = 0,$$
$$r(\{1\}) = 1,$$
$$r(\{0,1\}) = 1$$

Then $\langle \Sigma, r \rangle$ is a space with a risk for the system S, which is described by a group of two factors.

2. Let the system S is described by a finite group G_S, which is consist of three factors. Then $G_S \cong Z_3$ is a finite cyclic group of the order 3 with the following Cayley table

	0	1	2
0	0	1	2
1	1	2	0
2	2	0	1

$\Sigma = \langle \{\emptyset, \{0\}, \{1\}, \{2\}, \{0,1\}, \{0,2\}, \{1,2\}, \{0,1,2\}\}, \cup, \cap, \backslash, \prime \rangle$ is a σ-algebra for system S risks' description.

Let

$$\begin{aligned} r(\emptyset) &= 0, \\ r(\{0\}) &= 0, \\ r(\{1\}) &= 1, \\ r(\{2\}) &= 0 \\ r(\{0,1\}) &= 0 \\ r(\{0,2\}) &= 0 \\ r(\{1,2\}) &= 1 \\ r(\{0,1,2\}) &= 1 \end{aligned}$$

Then $\langle \Sigma, r \rangle$ is a space with a risk for the system S, which is described by a group of three factors.

We shall consider such systems, which are represented by finite groups of factors. Then σ-algebra for description risks of the system S will be finite.

Step 1. Define the purpose of the regulation.
Step 2. Define the strategy of regulation.
Step 3. Construct a space with the risk $\langle \Sigma, r \rangle$ for a system S described by a group of factors G_S. Let $G_S = \{e, a_1, .., a_{n-1}\}$. Let's consider the set $P(G_S)$ of all subsets of the set G_S. Let $\Sigma \subseteq P(G_S)$ and we shall consider an algebra $\mathbf{\Sigma} = \langle \Sigma, \cup, \cap, \setminus, ' \rangle$. An algebra $\mathbf{\Sigma}$ is a finite σ-algebra. Let's define the risk function on $\mathbf{\Sigma}$ to be equal to the restriction $r \lceil \Sigma$, where $r : P(G_S) \to R$ where $r : P(G_S) \to R$ is a risk function on a finite σ-algebra $\boldsymbol{P(G_S)} = \langle P(G_S), \cup, \cap, \setminus, ' \rangle$.
Step 4. Let's prove the existence of r. For this purpose it is enough to construct, double-digit risk function for example. Let $G_S = \{e, a_1, .., a_{n-1}\}$ consists of n elements. Then $P(G_S)$ consists of 2^n elements. Besides it, $\boldsymbol{P(G_S)} = \langle P(G_S), \cup, \cap, \setminus, ' \rangle$ is a field of subsets of a non-empty set G_S, consisting of 2^n elements. According to the work of Sikorski [9] $\boldsymbol{P(G_S)} = \langle P(G_S), \cup, \cap, \setminus, ' \rangle$ has a two-valued measure and a two-valued homomorphism, and hence an epimorphic image of two elements, that is, there is an epimorphism r: $\boldsymbol{P(G_S)} \to \langle \{0,1\}, \cdot \rangle$ or the risk function $r : P(G_S) \to \{0,1\} \subseteq R$. We note that in this particular case the risk function is a two-valued measure [9].
Step 5. We consider $\mathbf{\Sigma}^1 = \langle \{A | A \in \Sigma, r(A) = 1\}, \cap \rangle$ for a system S describing by a group of factors G_S, and construct the lattice of subsets for $\mathbf{\Sigma}^1 \cap \boldsymbol{P(G_S)}$. Then $\Sigma^1 \cap P(G_S)$ is a partially ordered set towards the binary relation of inclusion: $\langle \Sigma^1 \cap P(G_S), \subseteq \rangle$. After that we look for the minimal elements of $\langle \Sigma^1 \cap P(G_S), \subseteq \rangle$. Let $M_1 = \{A_i | i \in I_1\}$ be a set of minimal elements of $\langle \Sigma^1 \cap P(G_S), \subseteq \rangle$. After that we construct "risk preference relation" that is the partial order relation ρ on the set M_1.
Then we exclude the most undesirable elements from the set $M_1 = \langle \{A_i | i \in I_1\} \rangle$ with the help of directed influence on the system S.
Step 6. We repeat step 5 for set $\Sigma^1 \setminus M_1$.

After a finite number of steps the process will be completed, since the group of factors G_S is finite, and we shall neutralize all systemic risks of system S, described by a finite group of factors.

7.8.5 Some Properties of Risk. Examples

Example 1 Financial smart systems' risks. Let's formalize the term "system" as a set of elements structured by means of links into subsystems of different levels to achieve the goals of functioning of the system (goals' functions). We shall consider the system's risk. If we consider the risk of the system as a system of factors, in interaction with external and internal conditions of functioning, which determines the real events of the environment of the functioning of the system under study, then it has the following difference from all other systems: the purpose of its functioning is to ensure the nonequilibrium state of the functioning environment, since it is known from the theory of dynamical systems that in the absorbing or in the limiting state (equilibrium) the system ceases to exist. The risk of the system will be called systemic risk.

Theorem 7.13 (theorem about the description of systemic risk)
Systemic risk is determined by no more than two combinations of factors that determine the system's risk subsystem.

Proof Let the financial risk of a system be determined by a set of factors $\{a_1, a_2, \ldots, a_n\}$. We shall assume that the factors are independent, and that there are statistical observations showing the dependence of each factor on time

$$f_{a_i}(t), i = 1, \ldots, n$$

or there are density distributions of factors. Let's consider the free groups of factors

$$F_n = \langle a_1, a_2, \ldots, a_n \rangle$$

of the rank $\boldsymbol{n}$ with the generators $\boldsymbol{a}_1, \boldsymbol{a}_2, \ldots, \boldsymbol{a}_n$. Since any free group of finite rank is embeddable in a free group of countable rank, and the commutant $[\boldsymbol{F}_2, \boldsymbol{F}_2]$ of a free group $\boldsymbol{F}_2$ of the rank 2 is a free group of countable rank we can embed.

$\boldsymbol{F}_n = \langle \boldsymbol{a}_1, \boldsymbol{a}_2, \ldots, \boldsymbol{a}_n \rangle$ into $\boldsymbol{F}_\infty = [\boldsymbol{F}_n, \boldsymbol{F}_n]$ which is a free group of countable rank. After that we embed $\boldsymbol{F}_\infty = [\boldsymbol{F}_n, \boldsymbol{F}_n]$ into commutant $[\boldsymbol{F}_2, \boldsymbol{F}_2]$ of a free group $\boldsymbol{F}_2$ of the rank 2. So there exist two words $\boldsymbol{w}_1(\boldsymbol{a}_1, \boldsymbol{a}_2, \ldots, \boldsymbol{a}_n)$ and $\boldsymbol{w}_2(\boldsymbol{a}_1, \boldsymbol{a}_2, \ldots, \boldsymbol{a}_n)$ such that

$F_n = \langle a_1, a_2, \ldots, a_n \rangle$ is embedded into $F_2 = \langle w_1(a_1, a_2, \ldots, a_n), w_2(a_1, a_2, \ldots, a_n) \rangle$.

Let's suppose now that factors $a_1, a_2, \ldots, a_n$, which define the systemic risk of the system S, are distributed according to the same law with a probability density f, that is function f is given in the interval $(a, b) \subseteq (-\infty, +\infty)$, is continuous and lies in the strip $[0; 1]$. It means that for every $x \in (a, b) \subseteq (-\infty, +\infty)$ the inequality

$0 \leq f(x) \leq 1$ is true. Let's assume that factors $a_1, a_2, \ldots, a_n$ form a complete group of events which are independent in the aggregate at any time t.

We shall also assume that f satisfies the following conditions:

$$(\forall x, y \in (-\infty, +\infty))(f(x \cdot y) = f(x) + f(y)) \tag{7.18}$$

That is f is a homomorphism of an associative semigroup $\langle R, \cdot \rangle$ into the additive semigroup $\langle R^+ \cup \{0\}, + \rangle$.

Let's consider now $F_2 = \langle w_1(f(a_1), f(a_2), \ldots, f(a_n)), w_2(f(a_1), f(a_2), \ldots, f(a_n)) \rangle$. From (7.15) we get

$$\begin{aligned} f(F_2) &= f(\langle w_1(a_1, a_2, \ldots, a_n), w_2(a_1, a_2, \ldots, a_n) \rangle) \\ &= \langle w_1(f(a_1), f(a_2), \ldots, f(a_n)), w_2(f(a_1), f(a_2), \ldots, f(a_n)) \rangle \\ &\preccurlyeq \langle R^+ \cup \{0\}, +. \rangle \end{aligned}$$

Let $R_f = \langle w_1(f(a_1), f(a_2), \ldots, f(a_n)), w_2(f(a_1), f(a_2), \ldots, f(a_n)) \rangle \preccurlyeq \langle R, \cdot \rangle$, and f displays R_f into $f(F_2)$.

Then $\int_{-\infty}^{+\infty} f(F_2) \mathrm{d}f$ is a measure of systemic risk of the system S.

Note 7.14

For example, the exponential function satisfies the equality (7.18).

Example 2 Risks of P-effective and P-innovative systems.

In Chaps. 5 and 6 we have considered concepts of P-effective and P-innovative systems. Such systems are risk-free.

References

1. Prigozhyn I., Stengers, I.: Order from chaos. A new dialogue between man and nature. Progress, Moscow (1986). (in Russian)
2. Uskov, A.V., Serdyukova, N.A., Serdyukov, V.I., Heinemann, C., Byerly, A.: Multi objective optimization of VPN design by linear programming with risks models. Int. J. Knowl.-Based Intell. Eng. Syst. **20**(3), 175–188 (2016)
3. Venttsel, E.S.: Operations Research: Tasks, Principles, Methodology. Science, the Main Edition of Physical and Mathematical Literature, Moscow (1980). (in Russian)
4. LNCS Homepage http://life-prog.ru/1_29279_differentsialnie-uravneniya-kolmogorova.html
5. Zacharov, V., Kravtsov, A., Priymenko, C.: Generalizations of Bernstein polynomials in problems of multiobjective optimization. Sci. World **1**(6), 2–28 (2014)
6. Fikhtengol'ts, G.M.: A Course of Differential and Integral Calculus, vol. 2. Fizmatlit, Moscow (2001). (in Russian)
7. LNCS Homepage http://www.ideationtriz.com
8. Gnedenko, B., Ushakov, I., Falk, J. (eds.): Probabilistic Reliability Engineering. A Wiley-Interscience Publication, New York, Chichester, Brisbane, Toronto, Singapore (1995)
9. Sikorski, R.: Boolean Algebras. Springer, Berlin, Gottingen, Heidelberg, New York (1964)

Chapter 8
The Transition from an Infinite Model of Factors that Determine the System to a Finite Model. The Model of Algebraic Formalization of Risks of Changing the Scenarios of the Long-Term Development of a Smart System of Six Factors on the Example of a Smart University

Abstract In this chapter we shall consider the question of how from an infinite model of factors that determined the system S one can go to the finite model of factors G_S which determine the system S. A list of necessary information from the Finite Groups Theory, useful in the study of certain features of the functioning of the smart system, is given in addition. Table 8.1 in which some system's properties are classified is constructed on this basis by the models of finite groups of factors determining the system. The question about risk modeling in a smart university will also be considered in this chapter. The model of an algebraic formalization of six factors of the risks of changes in long-term period of a development of the smart system is constructed on the example of the smart university. The algorithm of search of points of regulation of the closed associative system's functioning on the example of the model consisting of six factors is shown in this chapter too.

Keywords Algebraic Formalization · Finite Group Theory · Risk

8.1 The Transition from an Infinite Model of Factors that Determine the System to a Finite Model of the System

Let us consider the set of factors $G = \{a_1, \ldots, a_n\}$ that define the system S. If G is not closed under the operation of factors' composition then we shall consider a free group $\boldsymbol{F_n}$ of the rank n with free generators $a_1, \ldots, a_n$. According to Chap. 6, definition 6.6, a free group $\boldsymbol{F_n}$ is a model of the system S. A free group $\boldsymbol{F_n}$ is a pseudo-innovation model of the system S as $\boldsymbol{F_n}$ is a projective group. A free group

N. Serdyukova and V. Serdyukov, *Algebraic Formalization of Smart Systems*, Smart Innovation, Systems and Technologies 91,
https://doi.org/10.1007/978-3-319-77051-2_8

$\boldsymbol{F}_n$ has an empty set of defining relations, so $\boldsymbol{F}_n$ does not reflect the connections of the system S. Now let's suppose that the operation of factors' composition is a commutative one. Then the factor-group $\boldsymbol{F}_n/[\boldsymbol{F}_n, \boldsymbol{F}_n]$ is a free abelian group of the rank n which is a model of the system S. A factor-group of the group $\boldsymbol{F}_n$ by its some normal subgroup $\boldsymbol{N} \trianglelefteq \boldsymbol{F}_n$ can be considered as a model of the system with defining factors $a_1, \ldots, a_n$, in which connections between factors is defined by the normal subgroup $\boldsymbol{N}$ of a group $\boldsymbol{F}_n$.

This construction helps to construct the finite model $\boldsymbol{G}_S$ of the system S in the form of a finite group of factors determined the system S. In fact to construct a Cayley table for the group $\boldsymbol{G}_S$ one can act in two following ways:

(1) To use combinatorial methods. One should search the defining relations of the model $\boldsymbol{G}_S$ with the help of simple enumeration.
(2) To make a qualitative analysis of the factors which determined the system and on this basis to explain the relationships between them. After that to construct a normal subgroup $\boldsymbol{N} \trianglelefteq \boldsymbol{F}_n$, which determines the connections between the factors $\{a_1, \ldots, a_n\}$, or in other words to describe the relations of the system S with the help of the model $\boldsymbol{G}_S$.

The following well-known facts from the Finite Groups Theory [1, 2], make it possible to pick out a number of properties of the model $\boldsymbol{G}_S$ of factors which determined the system S, see Table 8.1.

8.2 The Necessary Information from the Finite Groups Theory Useful in the Study of Some Features of the System's Functioning

Let us give now some known and interesting facts from the theory of finite groups [1, 2], that can be useful in studying such properties of the system as the presence of synergistic effects, the number of possible variants of forecasts for the development of the system, the stability properties of the system, and so on.

1. For groups of order $n < 60$ simple groups are precisely the cyclic groups $\boldsymbol{Z}_n$ where n is a prime number [1]. Let us note that for the purposes of mathematical modeling it is usually sufficient to have no more than 60 variables.
2. Theorem. Every group $\boldsymbol{G}$ of the order p^2, where p is a prime number, is an abelian group.
3. From the Silov's theorem we have that there are exactly two up to isomorphism groups of the order pq provided that $p < q$. One of them is an abelian group and the other is non-abelian one. The non-abelian group of the order pq exists if and only if the condition $p \equiv 1(modq)$ takes place.
4. An abelian simple group is a cyclic group of the prime order.

5. The Galois theorem. The even permutation group A_n is a simple group for $n \geq 5$.
6. Theorem (Faith, Thompson). The odd-order group is solvable.
7. The commutant of a group $\boldsymbol{G}$ is its subgroup $[G, G]$ which is generated by the commutators $[a, b] = a^{-1}b^{-1}ab$, where $a, b \in G$. A group $\boldsymbol{G}$ is called a soluble group if for some finite number k its subgroup $\boldsymbol{G}^k = E$, where $\boldsymbol{G}^k = [\boldsymbol{G}^{k-1}, \boldsymbol{G}^{k-1}] = [[\boldsymbol{G}^{k-2}, \boldsymbol{G}^{k-2}], [\boldsymbol{G}^{k-2}, \boldsymbol{G}^{k-2}]] = \ldots = [\ldots[\boldsymbol{G}, \boldsymbol{G}], [\boldsymbol{G}, \boldsymbol{G}], \ldots]$. If for any finite k we have $\boldsymbol{G}^k \neq E$, then the group $\boldsymbol{G}$ is called unsoluble one. Soluble simple groups are just cyclic groups of prime orders.
8. Theorem. Any p-group is a soluble group.
9. A group of order pq, where p and q are distinct is a soluble group.
10. A finite group is a soluble one if and only if there is a sequence of normal subgroups

$$G = H_0 \trianglerighteq H_1 \trianglerighteq \ldots \trianglerighteq H_k = E$$

such that every factor-group H_m/H_{m-1} is a cyclic group of a prime order.

11. The only groups with exactly two automorphisms are cyclic groups of orders 3, 4 and 6.
12. All finite groups are Hopfian groups. Recall that a group G is said to be non-Hopfian if it contains a non-trivial normal subgroup $H \trianglelefteq G$ such that $G/H \cong G$. If such a subgroup $H \neq E$ does not exist, then G is said to be a Hopfian group.
13. There are at most finitely many non-isomorphic finite groups G such that $Aut(G) \cong H$ for a given finite group H. Herewith not every group can be represented as a group of automorphisms of some group. For example, there is no finite or infinite group G such that $Aut(G) \cong Z_{2l+1}$ is a cyclic group of odd order. The symmetric groups of all permutations S_6 of degree 6, neither one of the non-trivial alternating groups A_n, except $A_8 \cong PSL(4, 2)$ and many other groups are not realized as a group of automorphism of a group, for example, the infinite cyclic group Z.
14. Dixon's Theorem. A cyclic group is a unique group of order n if and only if n and $\varphi(n)$ are mutually prime numbers, where $\varphi(n)$ is Euler's function, $\varphi(n)$ is equal to the number of natural numbers less than n and mutual prime with n, $\varphi(1) = 1$.
15. Orders of simple groups less than 1000000. There are exactly 56 non-abelian simple groups whose orders are less than 1000000. Most of these groups are projective special groups $PSL(2, q)$ of degree 2. Five of them are alternating groups A_5, A_6, A_7, A_8, A_9, three of which are isomorphic to $PSL(n, q)$. Five groups are sporadic ones:

    - three small Mathieu's groups M_{11}, M_{12}, M_{22},
    - two younger Janko's groups J_1, J_2.

All other groups, with the only exception are groups like Lee's classic series PSL_2, PSL_3, PSL_4, PSp_4, PSU_3, PSU_4. The only exception is the Suzuki's group $Sz(2^3)$.

All simple groups of order less than 1000000, except of two non-isomorphic groups $PSL(4,2)$ and $PSL(3,4)$ of order 20160 are uniquely determined by their order.

16. According to [2] a group G is called to be perfect if it coincides with its commutant that is if $[G,G]=G$. The term was introduced by Sophus Lee. The alternating group A_n, $n \geq 5$, is perfect. If G is non-abelian simple group, then $[G,G]=G$.
17. Ore's Theorem. If $n \geq 5$ then every element of group A_n is a commutator.
18. Ore's Problem. Is it true that every element of a non-abelian finite simple group is a commutator? For almost all finite simple groups it is known that the answer to this question is positive.

Note

In Chaps. 3 and 10 we have formulated and considered the notion of a finitely sustainable system. We find out from [1, 2], that there are exactly 56 models of finally sustainable closed systems with a feedback the number of elements of which does not exceed 1000000. All of these models except the two non-isomorphic groups $PSL(4,2)$ and $PSL(3,4)$ of order 20160 are uniquely determined by their order.

Table 8.1 Classification of system's properties by models of finite groups of factors that determine the system

Model $\boldsymbol{G_S}$	$\boldsymbol{G_S}$—is a simple group of the order less than 60. Then we have $\boldsymbol{G_S} \cong \boldsymbol{Z_n}$, where n is a prime number, $\boldsymbol{Z_n}$ is an abelian cyclic group of the order n
The number of elements of $\boldsymbol{G_S}$ (The number of factors of $\boldsymbol{G_S}$ that determine the system $\boldsymbol{S}$)	The order of group n is less than 60
The number of final states of the system $\boldsymbol{S}$	1
The number of possible synergetic effects of the system $\boldsymbol{S}$	There are no synergetic effects (The synergetic effects cannot be described by the model $\boldsymbol{G_S} \cong \boldsymbol{Z_n}$)
The description of the group $Hom(\boldsymbol{G_S}, GL(n,R))$ (tensor estimation of the system efficiency $\boldsymbol{S}$)	$Hom(\boldsymbol{Z_n}, GL(n,R))$

The next question that we shall consider in this chapter is the question about risks' modeling in a smart university.

8.3 The Model of an Algebraic Formalization of Risks of Changing the Scenarios of the Long-Term Development of a Smart System of Six Factors on the Example of a Smart University

The main idea of this section, which unites the further presentation of Chap. 8, is to show that, with the correct and timely regulation the process of the system's functioning, it becomes a smart system in the sense of the optimal system on the selected smart criteria, or, in other words, a smart optimal system.

In the proposed model of algebraic formalization of risks of changing the scenarios for the long-term development of a smart system of six factors, the risk of formalization's change from the symmetric scenario to the cyclic scenario, and the tensor index of the effectiveness of the system performance on specific indicators, can be calculated by the algorithm proposed below, for example, for the Russian economy [12], and for a smart university system. The main characteristics of smart university are described in [3–11].

The conceptual model of SmU—*CM-SmU*—can be described as follows [3-11].

Definition 8.1
Smart University is described as *n*-tuple of *n* elements, $n \in N$ which can be chosen from the following main sets:

$$\begin{aligned}\text{CM-SmU} = <&\{\text{SmU\_STUDENTS}\}, \{\text{Sm\_CURRICULA}\},\\ &\{\text{Sm\_FACULTY}\}, \{\text{Sm\_PEDAGOGY}\},\\ &\{\text{Sm\_CLASSES}\}, \{\text{Sm\_SOFTWARE}\},\\ &\{\text{Sm\_HARDWARE}\}, \{\text{Sm\_TECHNOLOGY}\},\\ &\{\text{Sm\_RESOURCES}\}>\end{aligned} \tag{8.1}$$

where:

SmU_STUDENTS	a set of types of students at SmU (local, remote, special types of students, undergraduate, graduate, etc.);
Sm_CURRICULA	a set of smart programs of study and smart courses at SmU —those that can change its structure in accordance with types of students in those programs and courses;
Sm_FACULTY	a set of faculty (instructors) at SmU who are trained and actively use smart systems, smart technology, smart boards, smart classrooms, etc.,
Sm_PEDAGOGY	a set of pedagogical styles (strategies) used at SmU; in general case, it may include a great variety of teaching and learning styles such as collaborative (local and remote students together), online learning, learning-by-doing, project-based learning, games-based learning, etc.,

Sm_CLASSES a set of smart classrooms, smart labs, smart departments and offices at SmU;

Sm_SOFTWARE a set of university-wide smart software systems at SmU; in general case, it may include a great variety of software systems for smart classrooms, smart labs, university-wide learning management systems, security systems, identification systems, etc.,

Sm_HARDWARE a set of university-wide smart hardware systems, devices, equipment and technologies used in SmU software systems at SmU; in general case, it may include a great variety of sensors, video cameras, smart interactive boards, etc.,

Sm_TECHNOLOGY a set of university-wide technologies to facilitate main functions of SmU; in general case it may include sensor technology, RFID technology, ambient intelligence technology, Internet-of-Things technology, wireless sensor networks, security and safety technology, etc.,

SmU_RESOURCES a set of various resources of SmU (financial, technological, human, etc.)

In general case, SmU may have multiple additional minor sets; however, for the purpose of this research project we will limit a number of SmU sets as presented in (1). The designated SmU components (below-factors) may have a great variety of relations (below-connections), for example, *SmU_students-to-Sm_faculty, Sm-faculty-to-Sm_software, Sm_curricula-to-Sm-pedagogy*, etc.

8.3.1 Risk Modelling in a Smart University

Let's return to tables 6.19 and 6.20 of the education system's effectiveness, which we examined in Chap. 6, and let's continue to study the process of decomposition of the education system. Using the same method as in Chap. 6, we shall consider the interconnections and mutual influence of the subsystems that we singled out of the smart university on each other.

It is necessary to recall a notion of system's final state.

Definition 8.2 Let $S(t) = S'$ be a transition of system S at the moment t into a new state S'. A set of all system S states, which occur during its operation, is denoted as $\{\ S(t_i)|i\epsilon I\}$, where t_i is a point of time for every $i\epsilon I$. During its operation the system S converts itself into system S_f at the point of time t_f. After that time a system stops any changes, or, in other words, a group of factors $G_S = G_S(t_f)$ that represent system S stop to change. Let call $G_S = G_S(t_f)$ as the final state of system S.

Since we consider models of systems with a finite number of factors that determine the system, one can compute the number of final states of the system as follows. Let the number of elements of the group $\boldsymbol{G_S}$ that determine the system S is equal to $|G_S| = n$. Then the number of final states of a system is equal to the number of non-isomorphic groups of the order n.

Let consider a system of education from this point of view. In general case, a system of education may contain a huge number of subsystems each of which can represent one or a part or several SmU components that are described in [3–11]. However, for the purpose of this research project and in order to simplify an explanation of developed methodology, below we will take into a consideration only the following sub-systems of system S:

(1) S_1—a subsystem of learning content or knowledge (let us call it informational subsystem); it corresponds exactly to the set {Sm_CURRICULA} in (8.1);
(2) S_2—a subsystem of pedagogy or methodological and methodical subsystem (let us call it adaptive subsystem); it corresponds exactly to the (Sm_PEDAGOGY} set in (8.1);
(3) S_3—a subsystem of students (let us call it target subsystem); it corresponds exactly to the {SmU_STUDENTS} set in (8.1);
(4) S_4—a financial subsystem (let us call it providing subsystem); it corresponds to a part of {SmU_RESOURCES}set in (Eq. 8.1).

The interaction of these subsystems and the connection between the considered subsystems of system S can be described as follows:

Stage 1.
S_2 affects S_1; an improvement of learning content takes place; the outcome of this action is designated as $S_2(S_1,)$.
Stage 2.
$S_2(S_1)$ affects S_3; as a result, there is a transformation of S_3 into $(S_2(S_1))(S_3)$.

The overall goal of a system of education is to find optimal final state $(S_2(S_1))$ (S_3) for a system in accordance with the following selected criteria:

(1) a completeness, necessity and sufficiency of an information subsystem S_1;
(2) a clarity, simplicity and accessibility for a perception of an adapting subsystem S_2;
(3) a required level of availability of a target subsystem S_3;
(4) a sufficient level of support by providing subsystem S_4.

8.4 A Selection of Factors to Determine Long-Term Risks of a System

It is necessary to identify indicators s_i that will help us to determine the long-term risks of a system of education; in general case a system may have numerous indicators. However, for the purpose of this research project and in order to simplify an explanation of developed methodology, below we will take into a consideration only the following indicators s_i, $i = 1,\ldots,6$:

(1) s_1—an indicator of system's financial status (for example, university's total assets that may include tuition and fees, government grants and appropriations, investments, contributions, property, equipment, etc.);
(2) s_2—an indicator of system's target subsystem (for example, a total number of students of various types such as undergraduate students, graduate students, life-long learners, etc.);
(3) s_3—an indicator of systems' adaptive subsystem (as the indicator one can chose the level of the development of methodological and methodical support of the knowledge system, it can be measured by the frequency of using the methodological support of the knowledge system by the personnel of higher qualification and by the target audience);
(4) s_4—an indicator of the required level of availability for target audience (test control of knowledge of the target audience and its monitoring);
(5) s_5—an indicator of sufficiency of the financial support for knowledge subsystem (for example, external funds for a system of education, or funds for special projects—disabled students, life-long learners, etc.);
(6) s_6—an indicator of accessibility of a system of education to students (for example, TOEFL level for international students, tuition fee for a credit hour, etc.).

Under the assumption that such a system which consists of 6 factors is closed and associative one we should choose the indicators of the factors and the position of the factors in the Cayley table of group in such a way that the table would have a physical sense we obtain, for example, the following two possible final scenario of the forecast:

Scenario# 1 corresponds to the cyclic group of the order 6, i.e. group Z_6 (Fig. 8.1):
Scenario# 2 corresponds to the symmetric group of permutations of the degree 3, i.e. group S_3 (Fig. 8.2).

Let's show how to use the tables. The analysis of scenarios 1 and 2 gives us the following.

Fig. 8.1 Cyclic group of the order 6

∘	s_1	s_2	s_3	s_4	s_5	s_6
s_1	s_1	s_2	s_3	s_4	s_5	s_6
s_2	s_2	s_3	s_4	s_5	s_6	s_1
s_3	s_3	s_4	s_5	s_6	s_1	s_2
s_4	s_4	s_5	s_6	s_1	s_2	s_3
s_5	s_5	s_6	s_1	s_2	s_3	s_4
s_6	s_6	s_1	s_2	s_3	s_4	s_5

Fig. 8.2 Symmetric group of the degree 3

$\circ$	s_1	s_2	s_3	s_4	s_5	s_6
s_1	s_1	s_2	s_3	s_4	s_5	s_6
s_2	s_2	s_1	s_6	s_5	s_4	s_3
s_3	s_3	s_5	s_1	s_6	s_2	s_4
s_4	s_4	s_6	s_5	s_1	s_3	s_2
s_5	s_5	s_3	s_4	s_2	s_6	s_1
s_6	s_6	s_4	s_2	s_3	s_1	s_5

The equation $s_i{}^\circ s_j = s_k$ shows that the changes of factors s_i and s_j leads to the change of s_k.

The difference between scenarios. Every table contains 36 squares, while searching non-identical squares it makes sense to consider squares with coordinates (i,j), where $i,j \geq 2,\ \ i,j = 1,\ldots,6$. These are the following 17 squares: diagonal squares with coordinates (2,2), (3,3), (5,5) and non diagonal squares with coordinates (3,2), (4,2), (5,2), (6,2), (2,3), (4,3), (5,3), (2,5), (3,5), (4,5), (2,6), (3,6), (4,6), (5,6). The rest 19 squares in the 1st and 2nd scenarios are the same.

In the case of realization of each of scenario one has:

According to the first scenario the indicator s_3 reacts to the growth (change) of the indicator s_2, and according to the second scenario the indicator s_1 reacts to the growth (change) of the indicator s_2, and the indicator s_3 does not react.

According to the first scenario the indicator s_5 reacts to the growth (change) of the indicator s_3, and according to the second scenario the indicator s_1 reacts to the growth (change) of the indicator s_3, and the indicator s_5 does not react.

According to both the first scenario and the second scenario the indicator s_1 reacts to the growth (change) of the indicator s_4.

According to the first scenario the indicator s_3 reacts to the growth (change) of the indicator s_5, and according to the second scenario the indicator s_6 reacts to the growth (change) of the indicator s_5, and the indicator s_3 does not react.

According to both the first scenario and the second scenario the indicator s_5 reacts to the growth (change) of the indicator s_6.

According to the first scenario the indicator s_4 reacts to the interaction of indicators s_2 and s_3. According to the second scenario the indicator s_6 reacts to the interaction of indicators s_2 and s_3.

According to the first scenario the indicator s_5 reacts to the interaction of indicators s_2 and s_4. According to the second scenario the indicator s_5 also reacts to the interaction of indicators s_2 and s_4.

According to the first scenario the indicator s_1 reacts to the interaction of indicators s_2 and s_6. According to the second scenario the indicator s_3 reacts to the interaction of indicators s_2 and s_6.

According to the first scenario the indicator s_5 reacts to the interaction of indicators s_4 and s_2. According to the second scenario the indicator s_6 reacts to the interaction of indicators s_4 and s_2 etc.

Now let's determine the possible points (in time) of process control. At the points of difference of the scenarios listed above, one scenario of the functioning of the system may change to another.

The monitoring of the numerical values of the indicators $\{s_i | i = 1, \ldots, 6\}$ is necessary to identify possible change of scenario of development in the system's functioning.

We shall use the following algorithm to determine possible control points for the functioning of a closed associative system.

8.4.1 *Algorithm of Search the Points of Regulation of Functioning of the Closed Associative System on an Example of the Model Consisting of Six Factors*

1. Looking for all of the points of difference for each of the scenarios.
2. Possible risks are determined by the difference points in the Cayley tables of scenarios 1 and 2.
3. Let $s_i{}^\circ s_j = s_k$ in accordance to the first scenario of development of a system, and $s_i{}^\circ s_j = s_m$ in accordance to the second scenario of development of a system. If at the point $t = t_\propto$ the indicator s_k changes the trend and the indicator s_m does not change its trend then the first scenario takes place. In our case these are the following points:

   - according to the first scenario $s_4{}^\circ s_4 = s_1$;
   - according to the second scenario $s_4{}^\circ s_4 = s_1$; $s_3{}^\circ s_3 = s_1$; $s_2{}^\circ s_2 = s_1$.

Let us construct the schemes reflecting the dynamics of the development of the indicators $s_i,\ \ i = 1, \ldots, 6$ according to the monitoring of the numerical values of the indicators.

The Cayley tables show that in the first scenario, the indicator s_3 does not change the trend of dynamics depending on the change in the tendency of the indicator s_1, and on the second-changes.

If the second scenario for the development of the S system is undesirable, and the monitoring of the indicators $s_i,\ \ i = 1, \ldots, 6$ in the time period $[t, t + \Delta t]$ shows, for example, the dependence of changes in the s_3 values on the changes in the s_1, values, this means that the system develops on the second, undesirable, scenario.

In this case, measures to regulate the functioning of the S system are necessary. Similar algorithms can be constructed for any finite model $\boldsymbol{G_S}$ of the system S.

8.5 Conclusions. Future Steps

The goal of performed research was to develop a methodology of SmU modeling as a system based on an algebraic formalization of general systems' theory [3, 13, 14, 16], theory of algebraic systems [17], theory of groups [1, 2, 15], and generalizations of purities [16], and identify formal mathematical conditions for a system—in this case SmU—to become efficient and/or innovative.

References

1. Convay, J.H., Curtis, R.T., Parker, R.A., Wilson, R.A.: Atlas of Finite Groups. Claredon Press, Oxford (1985)
2. Vavilov, N.: Specific theory of groups. LNCS Homepage: http://pps.kaznu.kz/2/Main/FileShow/684223/89/124/3348/ (in Russian)
3. Serdyukova, N.A., Serdyukov, V.I., Slepov, V.A., Uskov, V.L., Ilyin, V.V.: A formal algebraic approach to modelling smart university as an efficient and innovative system. In: Smart Innovation, Systems and Technologies, vol. 59, pp. 83–96 (2016)
4. Coccoli, M., Guercio, A., Maresca, P., Stanganelli, L.: Smarter universities: a vision for the fast changing digital era. J. Visual Lang. Comput. **6**, 1003–1011 (2014)
5. Barnett, R.: The Future University. Routlege, New York (2012)
6. Temple, P. (ed.): Universities in the Knowledge Economy: Higher Education Organisation and Global Change. Routledge, New York (2011)
7. Tikhomirov, V.: The world on a way to smart education: new opportunities. Open Educ. **3**, 22–28 (2011)
8. Hilton, M.: In: Exploring the Intersection of Science Education and 21st Century Skills: A Workshop Summary. National Research Council (2010)
9. Richey, R.C., Klein, J.D., Tracey, M.W.: The Instructional Design Knowledge Base: Theory, Research, and Practice. Routledge, New York (2010)
10. Spector, J.: Conceptualizing the emerging field of smart learning environments. Smart Learn. Environ. (1) (2014)
11. Gamalel-Din, S.A.: Smart e-learning: a greater perspective: from the fourth to the fifth generation e-learning. Egypt. Inform. J. **11**, 39–48 (2010)
12. Ilyin, V.V., Serdyukova, N.A., Serdyukov, V.I.: Risks of long-term forecasts in the economy of the Russian Federation. Financial Analytics. Probl. Solutions **44**(278), 2–16 (2015). (in Russian)
13. Mesarovich, M., Takahara, Y.: General System Theory: Mathematical Foundations, Academic Press, New York, San Francisco, London, Mathematics in science and engineering, vol. 113 (1975)
14. Serdyukova, N.A.: Optimization of Tax System of Russia, Parts I and II, Budget and Treasury Academy, Rostov State Economic University (2002). (in Russian)
15. Kurosh, A.G.: Theory of Groups, Moscow, Nauka (1967) 648 p. (in Russian)
16. Serdyukova, N.A.: On generalizations of purities. Algebra Logic **30**(4), 432–456 (1991)
17. Malt'sev, A.I.: Algebraic Systems, Moscow, Nauka (1970) 392 p. (in Russian)

Chapter 9
Pro-*P*-Groups and Algebraically Closed Groups: Application to Smart Systems

Abstract In this chapter before we begin to examine pro-*P*-groups, we return again to the special case in which the factors affecting the system determine the group. In this case the system is a closed associative system with a feedback. Here we dwell briefly upon the modeling of "identical" factors with respect to the structure that act on the system. The question arises as to how all possible structures of connections between factors acting on the system can be described. We shall use the automorphism group of the group of factors that determine the system to this purpose. After that we recall the definition and basic information about algebraically compact groups that are necessary for the study of innovative and pseudo-innovative systems. Algebraically compact groups are in some way a generalization of divisible groups in two following directions: the first line (1) is distinguished as a direct summand from the group containing it, when (2) certain conditions are imposed on how the subgroup is contained in the overgroup. If a divisible group can be defined as a group distinguished as a direct summand from any group that contains it, then an algebraically compact group is a group distinguished as a direct summand from any group that contains it as a pure subgroup.

Keywords Algebraically compact groups · Direct and inverse limits
Pro-*P*-groups

9.1 Particular Case: Factors Affecting a System Determine a Group

Let S be a closed associative system with feedback.

In this case an algebra $\bar{A} = \langle A | \{ f_{\alpha}^{n_{\alpha}} | \alpha \in \Gamma \} \rangle$ with the main set of factors A and the set of operations $\{ f_{\alpha}^{n_{\alpha}} | \alpha \in \Gamma \}$, which describe the interaction of factors is a group $\bar{A} = \langle A |^{\circ}, \square^{-1}, e \rangle$, where $^{\circ}$—is the operation of factors' composition, that is sequential implementation of factors, $\square^{-1}$ is take-back operation, that is a realization of a feedback, e is a neutral factor. The property of a system to be a closed one is modeled by that the main set of factors A is closed with respect to the binary

N. Serdyukova and V. Serdyukov, *Algebraic Formalization of Smart Systems*,
Smart Innovation, Systems and Technologies 91,
https://doi.org/10.1007/978-3-319-77051-2_9

operation of the composition of the factors, the property of the system to be associative one is determined by the fact that the composition operation of the factors $\circ$ is associative one, and the feedback is modeled by the operation $\square^{-1}$ of taking an inverse element. Let's remind again, that under the algebra of factors of a system we shall understand an algebra $\bar{A} = \langle A | \{ f_\alpha^{n_\alpha} | \alpha \in \Gamma \} \rangle$ with a main set of factors A and the set of operations $\{ f_\alpha^{n_\alpha} | \alpha \in \Gamma \}$, that describe the interaction of factors.

The actions of the operations $\{ f_\alpha^{n_\alpha} | \alpha \in \Gamma \}$ on the set A show qualitative changes of the system under the influence of the factors affecting it. Let's recall further that:

A subalgebra $\bar{B} = \langle B | \{ f_\alpha^{n_\alpha} | \alpha \in \Gamma \} \rangle$ of an algebra $\bar{A} = \langle A | \{ f_\alpha^{n_\alpha} | \alpha \in \Gamma \} \rangle$ is called a *P*-pure subalgebra of an algebra $\bar{A}$, if every homomorphism $\bar{B} \xrightarrow{\alpha} \bar{C}$ of a subalgebra $\bar{B}$ of an algebra $\bar{A}$ into an algebra $\bar{C}$ of a signature $\{ f_\alpha^{n_\alpha} | \alpha \in \Gamma \}$, which is satisfied to predicate P, $P(\bar{C})$ is true, can be continued up to a homomorphism of an algebra $\bar{A} = \langle A | \{ f_\alpha^{n_\alpha} | \alpha \in \Gamma \} \rangle$ into an algebra $\bar{C} = \langle C | \{ f_\alpha^{n_\alpha} | \alpha \in \Gamma \} \rangle$, in such a way that the Diagram (9.1) will be a commutative one, that is

$$
\begin{array}{ccc}
0 \to \bar{B} = \langle B | \{ f_\alpha^{n_\alpha} | \alpha \in \Gamma \} \rangle & \xrightarrow{\varphi} & \bar{A} = \langle A | \{ f_\alpha^{n_\alpha} | \alpha \in \Gamma \} \rangle \\
\alpha \searrow & & \swarrow \beta \\
& \bar{C} = \langle C | \{ f_\alpha^{n_\alpha} | \alpha \in \Gamma \} \rangle &
\end{array}
\tag{9.1}
$$

the equality $\beta\varphi = \alpha$, where φ is an embedding $\bar{B} = \langle B | \{ f_\alpha^{n_\alpha} | \alpha \in \Gamma \} \rangle$ into $\bar{A} = \langle A | \{ f_\alpha^{n_\alpha} | \alpha \in \Gamma \} \rangle$, P is an unary predicate on the class of algebras of the signature $\{ f_\alpha^{n_\alpha} | \alpha \in \Gamma \}$, which singles out the class of subalgebras closed under subalgebras and factor-algebras,[1] is true, φ is called a *P*-pure embedding.

Purity is, in fact, a fractality (self-similarity) of connections. In other words, purity determines the quality of the links, in our case the quality of the bonds with respect to the condition *P*.

9.1.1 *The Meaning of the P-Pure Embeddings. Examples of P-Purities in the Class of All Groups*

For the classical definition of purities, [1], the Diagram (9.1) has the following meaning in the class of groups: the epimorphic images of $\bar{B}$ and $\bar{A}$ in the class of all finite groups are the same.

[1] The main operations of the same type of algebraic systems of the same signature will be denoted in each of the algebras in the same way.

For the P-purities the Diagram (9.1) has the following meaning: the epimorphic images of $\bar{B}$ and $\bar{A}$ in the class of all groups which satisfies P are the same.

We consider the following predicates as the main examples of the various conditions under which we shall study the functioning of the systems:

P singles out the class of all finite groups in the class of abelian groups, we obtain the usual purity of abelian groups;
P singles out the class of all abelian groups in the class of all groups;
P singles out the class of all finite groups in the class of all groups;
P distinguishes the variety in the class of all groups; that is the class closed with respect to subgroups, to homomorphic images and to Cartesian products, for example, the Burnside variety of all groups of exponent n, defined by the identity $x^n = 1$, the variety of nilpotent groups of nilpotency class not greater than n, solvable groups of length not greater than l, etc.

An important role in this case is played by "identical" factors with respect to the structure that act on the system. To clear up this we shall use the group of automorphisms of the factors that determine the system.

9.2 The Group of Automorphisms of the Group of Factors that Determine the System

Let us dwell briefly upon the modeling of "identical" factors with respect to the structure that act on the system. The question arises as to how all possible structures of connections between factors acting on the system can be described. We shall use the automorphism group of the group of factors that determine the system to this purpose.

Recall that the groups G and G' are isomorphic if a one-to-one mapping φ, that preserved the binary operation can be established between the elements of their main sets. An isomorphism of a group onto itself is called an automorphism of the group. The concept of isomorphism allows one to distinguish an algebraic binary operation as an object to study. Let $\bar{G} = \langle G|^{\circ}, \Box^{-1}, e \rangle$ be a group of factors which describe the system $\breve{G}$. Then the automorphism group $Aut(\bar{G})$ shows all possible structures of the links of factors acting on the system $\breve{G}$ in exactly the same way as $\bar{G}$.

9.2.1 *Background of the Issue. Basic Definitions and Theorems*

A natural generalization of algebraically compact groups and profinite groups are pro-P-groups, where P-is a predicate defined on the class of all groups and closed with respect to taking subgroups and factor groups. More generally, one can define pro-P-algebraic systems of signature Ω, where P is a predicate defined on the class of all algebraic systems of signature Ω, closed with respect to taking subsystems and factor systems.

We shall continue the study of P-innovative and P-pseudo-innovation systems with the help of this notion.

We recall the definition and basic information about algebraically compact groups that are necessary for the study of innovative and pseudo-innovative systems. Algebraically compact groups are in some way a generalization of divisible groups in two following directions: the first line (1) is distinguished as a direct summand from the group containing it, when (2) certain conditions are imposed on how the subgroup is contained in the overgroup. If a divisible group can be defined as a group distinguished as a direct summand from any group that contains it, then an algebraically compact group is a group distinguished as a direct summand from any group that contains it as a pure subgroup.

The following theorem, [1, p. 186], describes the characteristic properties of algebraically compact groups.

Theorem 9.1 [1, p. 186] *The following properties of the group G are equivalent*:

(1) *group G is pure injective,*
(2) *group G is algebraically compact,*
(3) group *G is a direct summand of the direct product of cocyclic groups,*
(4) *group G in the algebraic sense is a direct summand of a group admitting a compact topology,*
(5) *if every finite subsystem of a system of equations over G has a solution in the group G, then the whole system of equations is solvable in G.*

By means of a generalization to the non-commutative case, from (5) we obtain a definition of an algebraically closed group.

For the sequel we shall need the definition of direct and inverse spectra and their limits.

9.3 Direct and Inverse Spectra of Groups and Their Limits

Inverse Limits.

Let's follow [1] in the presentation of this material. We note that the concept of an inverse limit is dual with respect to the concept of a direct limit.

Definition 9.2 Let $\{G_i | i \in I\}$ be a system of groups, a partially ordered set of indices I is directed that is for every $i, j \in I$ there exists $k \in I$ such that $i \leq k, j \leq k$. Let for every pair of indices $i, j \in I$ such that $i \leq j$, a homomorphism $\pi_i^j : G_j \to G_i$ be given, at that the following conditions hold:

(1) π_i^i is the identity mapping of the group G_i for every $i \in I$,
(2) for every $i, j, k \in I$, such that $i \leq j \leq k$ the equality $\pi_i^j \cdot \pi_j^k = \pi_i^k$ takes place.

Then the system $\{G_i | i \in I, \pi_i^j\}$ is called an inverse spectrum.

The inverse (or projective) limit or simply the limit of the inverse spectrum $\{G_i | i \in I, \pi_i^j\}$ is called a subgroup $G^* = \lim_{\leftarrow} G_i$ of the direct product $G = \prod_{i \in I} G_i$, which consists of all vectors $g = (\ldots, g_i, \ldots, g_j, \ldots)$, such that $\pi_i^j g_j = g_i, i \leq j$.

We also need the following well-known facts concerning projective limits:

1. There exist such homomorphisms $\pi_i : G^* \to G_i, i \in I$, for which all the diagrams, where $i \leq j$

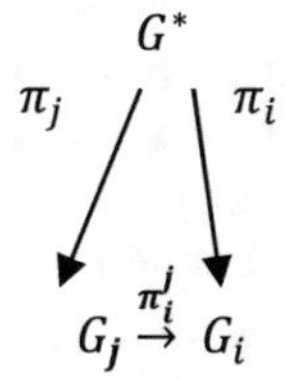

$$\begin{array}{ccc} & G^* & \\ {}^{\pi_j}\swarrow & & \searrow^{\pi_i} \\ G_j & \xrightarrow{\pi_i^j} & G_i \end{array} \tag{9.2}$$

are commutative.

2. The inverse limit G^* of the inverse spectrum $\{G_i | i \in I, \pi_i^j\}$ has the following property: if A is a group and $\sigma_i : A \to G_i$ are homomorphisms for which the diagrams

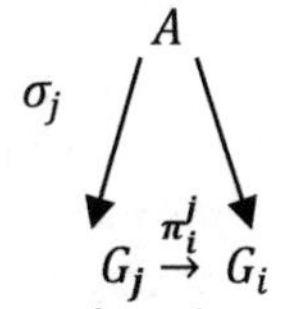

$$\begin{array}{ccc} & A & \\ {}^{\sigma_j}\swarrow & & \searrow \\ G_j & \xrightarrow{\pi_i^j} & G_i \end{array} \tag{9.3}$$

are commutative, then there exists unique homomorphism $\sigma : A \to G^*$, for which all diagrams

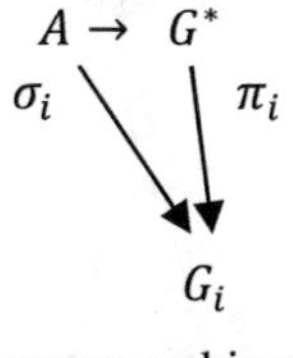

$$\begin{array}{ccc} A & \rightarrow & G^* \\ {}^{\sigma_i}\searrow & & \downarrow^{\pi_i} \\ & & G_i \end{array} \tag{9.4}$$

where $\pi_i : g \to g_i$ is a canonical homomorphism, that is it is a restriction of i-th coordinate projection of the group $\prod_{i \in I} G_i$ by the set G^*, are commutative. This property defines an inverse limit up to isomorphism.

3. If all the groups in the inverse spectrum $\{G_i | i \in I, \pi_i^j\}$ are (Hausdorff) topological groups, and all π_i^j are continuous homomorphisms, then the inverse limit G^* is a closed subgroup of a group $\prod_{i \in I} G_i$, if we consider the group $\prod_{i \in I} G_i$ endowed with the topology of the product, and the canonical homomorphisms $\pi_i : G^* \to G_i$ are continuous.

9.4 The Role of Profinite Groups in Algebra and Topology

9.4.1 Profinite Groups

Let us recall the definition and basic information about profinite groups necessary for studying the formalization of innovative and pseudo-innovative systems. A topological group that can be represented as a projective limit of finite groups is said to be a profinite one. The class of profinite groups coincides with the class of compact completely disconnected groups. The concept of the profinite group has been time and again generalized, see, for example, [2]. Thus, classes of pro-p-groups, where p is a prime number, pro-π-groups, where π is the set of prime numbers, pronilpotent groups, and pro-solvable groups were defined.

It is well known, [3, 4], that for every group G its profinite completion $\hat{G}$ can be constructed which is defined as the inverse limit of groups G/N, where N runs through the set of all normal subgroups of the group G of a finite index. The set of normal subgroups of a group G of finite index is partially ordered with respect to inclusion, thereby the inverse spectrum $\{G/N_i | i \in I, \pi_i^j\}$ of factor-groups of a group G by its normal subgroups of finite index:

$$\pi_i^j : G/G_j \to G/G_i$$

where $G_i \subseteq G_j$, $\pi_i^j(gG_j) = gG_i$ is defined. There exists unique homomorphism $\eta : G \to \hat{G}$, such that the image $\eta(G)$ of a group G under this homomorphism is dense in $\hat{G}$. A homomorphism η is an injective one if and only if the group G is a residually finite or finitely approximable one that is if $\bigcap_{i \in I} N_i = 1$. A homomorphism η satisfies the following condition: for any profinite group H and any homomorphism $f : G \to H$ there exists a unique continuous homomorphism $g : \hat{G} \to H$ such that the equality $f = g\eta$ takes place:

(9.5)

9.4.2 Profinite Completion

Given an arbitrary group G, there is a related profinite group $\hat{G}$, the profinite completion of G. It is defined as the inverse limit of the groups G/N, where N runs through the normal subgroups in G of finite index (these normal subgroups are partially ordered by inclusion, which translates into an inverse system of natural homomorphisms between the quotients). There is a natural homomorphism $\eta{:}G \to \hat{G}$, and the image $\eta(G)$ of G under this homomorphism is dense in $\hat{G}$. The homomorphism η is injective if and only if the group G is residually finite (i.-e., $\prod_{i \in I} N_i = 1$, where the intersection runs through all normal subgroups of finite index). The homomorphism η is characterized by the following universal property: given any profinite group H and any group homomorphism $f{:}G \to H$, there exists a unique continuous group homomorphism $g{:}\hat{G} \to H$ with $f = g\eta$.

9.4.3 P-*Finite Groups. Pro*-P-*Groups*

Let's define P-finite group where P is a predicate closed under taking subgroups and factor-groups.

Definition 9.3 A group G is called P-finite group, if $P(G)$ is true and G is a finite group.

Definition 9.4 A group is called P-profinite group, (or pro-P-group), if it is an inverse limit of a P-finite groups.

The question arises: for what predicates P class of pro-P-groups coincides with the class of inverse limits of groups G such that $P(G)$ is (that is if to remove the finiteness condition)? For example, if predicate P highlights the class of finite groups, then the class of pro-P-groups coincides with the class of inverse limits of groups G such that $P(G)$ is true.

It turns out that P-profinite groups are profinite groups satisfying the predicate P.

Let's define the P-topology.

To do this let's connect pro-P-groups with the inverse limits of P-injective groups. By analogy with [1], we formulate the definition of a P-algebraically compact group.

Definition 9.5 A group A is said to be P-algebraically compact if it is a retract of every group G, which contains it as a P-pure normal subgroup.

Let us recall the definition of a retract of a group.

Definition 9.6 A subgroup A of a group H is called a retract of H, if the following equivalent conditions are satisfied:

(1) There exists an endomorphism σ of a group H such that $\sigma^2 = \sigma$ and the image σ in H is equal to A, that is

$$(\exists \sigma \in End\, H)\big((\sigma^2 = \sigma) \wedge (Im \sigma = A)\big)$$

(2) There exists a normal subgroup N of a group H such that :$NA = H$ and $N \cap A = 1$, that is

$$(\exists N \trianglelefteq H)((NA = H) \wedge ((N \cap A) = 1)))$$

Theorem 9.7 *Let P be a predicate given on a class of groups which is closed under taking subgroups and factor-groups. Every group G, which is satisfied to predicate P can be embedded into pro- P-completion $\widehat{G_P}$ of a group G.*

The proof of this theorem follows the above arguments, which describe the imbedding of G into its profinite completion.

Theorem 9.8 *The following conditions are equivalent:*

(1) *a group A is P-pure injective;*
(2) *a group A is P-algebraically compact;*
(3) *a group A is a retract of an inverse limit of pro- P-groups.*

Proof Let us proof that $(1) \to (2)$. Let A be a P-pure subgroup of a group G. Then the short exact sequence

$$1 \to A \xrightarrow{i} G \xrightarrow{\pi} G/A \to 1$$

where i is a natural embedding, is a P-pure sequence and the diagram

$$\begin{array}{l} 1 \to A \xrightarrow{i} G \xrightarrow{\pi} G/A \to 1 \\ \quad\;\; \| \\ \quad\;\; A \end{array} \tag{9.6}$$

can be extended to a commutative one

$$\begin{array}{l} 1 \to A \xrightarrow{i} G \xrightarrow{\pi} G/A \to 1 \\ \quad\;\; \| \;\swarrow \vartheta \\ \quad\;\; A \end{array} \tag{9.7}$$

by a homomorphism $\vartheta{:}G \to A$, that is in such a way that $\vartheta i = 1_A$. It means that A is a retract of G. So, we have $(1) \to (2)$.

Let us show that $(2) \to (3)$. Let a group A be a P-algebraically compact group. First we shall proof that A is a retract of an inverse limit of pro-P-groups. Initially we embed A into its profinite completion

$$\hat{A} = \lim_{\leftarrow}\{A/A_i||A{:}A_i|<\infty, \pi_i^j{:}A/A_j \to A/A_i i \leq j; i,j \in I\}$$

This embedding is a pure injective one because the diagram

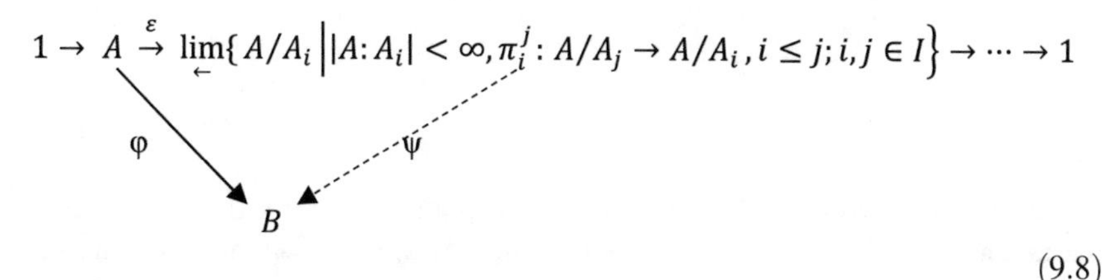

(9.8)

can be complemented up to a commutative one by a homomorphism

$$\psi{:}\lim_{\leftarrow}\{A/A_i|A{:}A_i||<\infty, \pi_i^j{:}A/A_j \to A/A_i, \\ i\leq j; i,j \in I, P(A/A_i), i \in I\} \to B$$

for which $\psi(\ldots, aA_i, \ldots, aA_j, \ldots) = \varphi(a)$. Let us note that $Im\varphi \leq B$, so as $P(B)$ is true and P is closed under taking subgroups, then $P(Im\varphi)$ is true and A has epimorphic images which satisfy predicate P.

Now embed A into it P-profinite completion $\widehat{A_P}$. Let's consider the inverse spectrum $\lim_{\leftarrow}\{A/A_i||A{:}A_i|<\infty, \pi_i^j{:}A/A_j \to A/A_i$is a natural projection, $i\leq j; i,j \in I, P(A/A_i), i \in I\}$. We should proof that $I \neq \emptyset$. It is so because above it was proved that A has epimorphic images which satisfy predicate P.

Let $\varepsilon{:}a \mapsto (\ldots, aA_i, \ldots, aA_j, \ldots)$ be an embedding

$$A \xrightarrow{\varepsilon} \lim_{\leftarrow}\{A/A_i||A{:}A_i|<\infty, \pi_i^j{:}A/A_j \to A/A_i \text{ is a natural projection}, \\ i\leq j; i,j \in I, P(A/A_i), i \in I\}.$$

Then ε is a P-pure embedding. In fact, for the diagram

$$1 \to A \xrightarrow{\varepsilon} \lim_{\leftarrow}\{A/A_i \,\big|\, |A{:}A_i| < \infty, \pi_i^j : A/A_j \to A/A_i, i \leq j; i,j \in I, P(A/A_i), i \in I\big\} \to \cdots \to 1$$

φ ψ B (9.9)

let's set $\psi(\ldots, aA_i, \ldots, aA_j, \ldots) = \varphi(a)$. Here $P(B)$ is true.

Let's consider the diagram

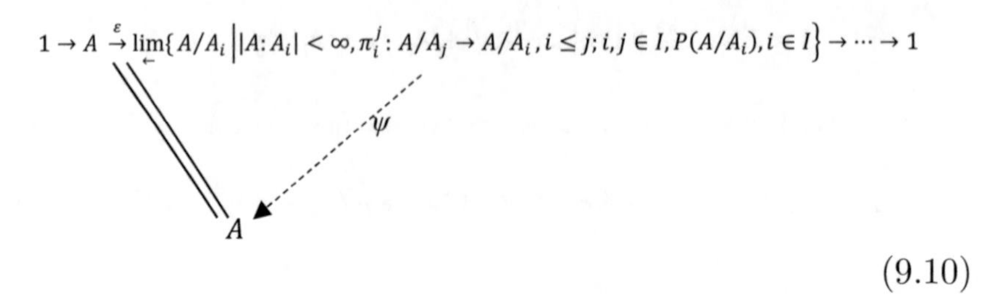

(9.10)

As we have assumed (2), then A is a P-pure injective group and thus the homomorphism ψ, that complements the diagram up to a commutative one there exists. So A is a retract of $\lim_{\leftarrow}\{A/A_i||A{:}A_i|<\infty, \pi_i^j{:}A/A_j \to A/A_i, i \le j; i,j \in I, P(A/A_i), i \in I\}$ and (3) is proved.

Let now A be a retract of an inverse limit of pro-P-groups. Let us prove that in this case A has a P-profinite completion $\widehat{A_P}$. The fact that A is a retract of an inverse limit of pro-P-groups means, that A is a subgroup of a group $\lim_{\leftarrow}\{G/G_i|\pi_i^j{:}G/G_j \to G/G_i, i \le j; i,j \in I\}$ such that $P(G/G_i)$ is true for $i \in I$ and G_i has a finite index in G for $i \in I$, and there exists an endomorphism σ of a group $\lim_{\leftarrow}\{G/G_i|\pi_i^j{:}G/G_j \to G/G_i, i \le j; i,j \in I\}$ such that $P(G/G_i)$ is true for $i \in I$ and G_i has a finite index in G for $i \in I$, wherein $\sigma^2 = \sigma$ and $Im\sigma = A$. Then $\widehat{A_P} = \lim_{\leftarrow}\{G \cap A/G_i \cap A|\pi_i^j{:}G \cap A/G_j \cap A \to G \cap A/G_i \cap A, i \le j; i,j \in I\}$ and $P(G \cap A/G_i \cap A)$ is true for $i \in I$, as P is closed under taking subgroups and factor-groups.

Let's now show that (3) $\to$ (1). Let a group A be a retract of inverse limit of pro-P-groups. We should proof that a group A is a P-pure injective one. Let we have a P-pure exact sequence

$$\begin{array}{l} 1 \to H \xrightarrow{\varepsilon} K \to K/H \to 1 \\ \varphi \downarrow \\ \quad A \end{array} \tag{9.11}$$

and $\varphi{:}H \to A$ be a homomorphism. Let's prove that there exists a homomorphism $\psi{:}K \to A$ such that $\psi\varepsilon = \varphi$. The fact that A is a retract of the inverse limit of pro-P-groups means that A is a subgroup of the group $\lim_{\leftarrow}\{G/G_i|\pi_i^j{:}G/G_j \to G/G_i, i \le j; i,j \in I\}$ such that $P(G/G_i)$ is true for $i \in I$ and G_i is a subgroup of a finite index of a group G for $i \in I$, and there exists an endomorphism σ of the group $\lim_{\leftarrow}\{G/G_i|\pi_i^j{:}G/G_j \to G/G_i, i \le j; i,j \in I\}$ such that $P(G/G_i)$ is true for $i \in I$ and G_i is a subgroup of a finite index of a group G for $i \in I$, wherein $\sigma^2 = \sigma$ and $Im\sigma = A$. As $\varprojlim\{G/G_i|\pi_i^j{:}G/G_j \to G/G_i, i \le j; i,j \in I\}$ is a P-pure injective group, then from this we obtain:

$$\lim_{\leftarrow}\{G/G_i|\pi_i^j, i \le j; i,j \in I\} \xrightarrow{\sigma} \lim_{\leftarrow}\{G/G_i|\pi_i^j, i \le j; i,j \in I\} \xrightarrow{\sigma} \lim_{\leftarrow}\{G/G_i|\pi_i^j, i \le j; i,j \in I\},$$

$$\sigma^2 = \sigma$$

$$Im\sigma = A$$

and there exists a homomorphism $\theta: K \to \lim_{\leftarrow}\{G/G_i|\pi_i^j, i \le j; i,j \in I\}$ such that the diagram

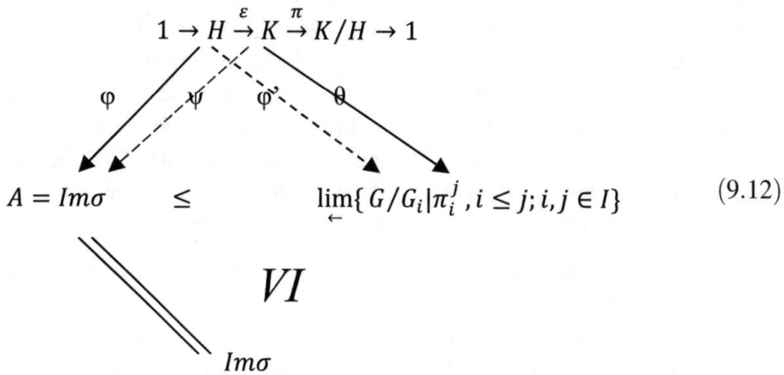

(9.12)

is commutative, and $\theta\varepsilon = \varphi'$, $\sigma\varphi = \varphi'$ and we can put $\psi = \theta$, because $Im\sigma = A$.

9.5 Predicates Defined by Systems of Equations on the Class of Groups

Let us now change the definition of the purities on predicates, eliminating the condition that the predicate P be closed with respect to taking the subalgebras, the Diagram (9.1):

Definition 9.9 A subalgebra $\bar{B} = \langle B|\{f_\alpha^{n_\alpha}|\alpha \in \Gamma\}\rangle$ of an algebra $\bar{A} = \langle A|\{f_\alpha^{n_\alpha}|\alpha \in \Gamma\}\rangle$ is called a *P*-pure subalgebra of an algebra $\bar{A}$, if every homomorphism $\bar{B} \xrightarrow{\alpha} \bar{C}$ from a subalgebra $\bar{B}$ of an algebra $\bar{A}$ into an algebra $\bar{C}$ of a signature $\{f_\alpha^{n_\alpha}|\alpha \in \Gamma\}$, such that $P(\bar{C})$ is true, where predicate *P* is sustainable with respect to factor-algebras, can be continued up to homomorphism from algebra $\bar{A} = \langle A|\{f_\alpha^{n_\alpha}|\alpha \in \Gamma\}\rangle$ into an algebra $\bar{C} = \langle C|\{f_\alpha^{n_\alpha}|\alpha \in \Gamma\}\rangle$, that is the Diagram (9.13) is commutative:

$$0 \to \bar{B} = \langle B|\{f_\alpha^{n_\alpha}|\alpha \in \Gamma\}\rangle \xrightarrow{\varphi} \bar{A} = \langle A|\{f_\alpha^{n_\alpha}|\alpha \in \Gamma\}\rangle$$

$$\alpha \searrow \qquad \swarrow \beta \qquad (9.13)$$

$$\bar{C} = \langle C|\{f_\alpha^{n_\alpha}|\alpha \in \Gamma\}\rangle$$

It means that $\beta\varphi = \alpha$, where φ is an embedding $\langle \bar{B} = B|\{f_\alpha^{n_\alpha}|\alpha \in \Gamma\}\rangle$ into $\langle \bar{A} = A|\{f_\alpha^{n_\alpha}|\alpha \in \Gamma\}\rangle, P$ is a predicate defined on the class of algebras of the signature $\{f_\alpha^{n_\alpha}|\alpha \in \Gamma\}$, highlights the class of subalgebras which is closed under taking factor-algebras φ is called a P-pure embedding. The main operations of the same type of algebraic systems of the same signature will be denoted in each of the algebras in the same way.

For such a definition of the P-purity there will be no duality, which is analogous to the duality described in [1].

Let us note that in the class of all groups in terms of the classical definition of purity, [1], Diagram (9.8) has the same meaning as the Diagram (9.1), that is: epimorphic images of $\bar{B}$ and $\bar{A}$ in the class of all finite groups are the same.

For P-purities the meaning of the Diagram (9.8) is exactly the same as the meaning of the Diagram (9.1) and runs as follows: epimorphic images of $\bar{B}$ and $\bar{A}$ in the class of all groups which satisfy condition P are the same. Thus, in this case, the condition of the closedness of the predicate P with respect to taking subgroups for this case is not essential, just as in the following examples:

P highlights the class of all finite groups in the class of Abelian groups, we get the usual purity in the class of Abelian groups;
P highlights the class of all Abelian groups the class of all groups;
P highlights the class of all finite groups in the class of all groups;
P highlights the variety in the class of all groups, that is a class closed with respect to subgroups, of homomorphic images and Cartesian products, for example, the Burnside variety of all groups of exponent n, defined by the identity $x^n = 1$, the variety of nilpotent groups of nilpotency class not greater than n, solvable groups of length not greater than l, etc.

In the framework of this new definition, the role of "identical" in structure factors acting on the system also saves: the automorphism group $Aut(\bar{G})$ shows all possible structures of the links of factors acting on the system $\breve{G}$ in exactly the same way as $\bar{G}$. Definition 9.9 was introduced by us in order to extend the analogy of P-algebraically compact groups with algebraically compact Abelian groups in the direction of Theorem 9.1, [1, p. 186], condition (5).

Let P be given by a system of equations of a signature $\Omega = \langle {}^\circ, \Box^{-1}, e\rangle$:

$$\wedge_{I \in i} w_i(x_1, x_2, \ldots, x_{n_i}) = e, \tag{9.14}$$

where $n_i \in N$.

It means that, $w_i(x_1, x_2, \ldots, x_{n_i})$ are words of the signature $\Omega = \langle {}^\circ, \Box^{-1}, e\rangle$ in the alphabet $\{x_1, x_2, \ldots, x_n, \ldots, e, x_1^{-1}, x_2^{-1}, \ldots, x_n^{-1}, \ldots\}$. Let $P \equiv \wedge_{i \in I} w_i(x_1, x_2, \ldots, x_{n_i}) = e$. Now let $\boldsymbol{G} = \langle G, {}^\circ, \Box^{-1}, e\rangle$ be a group which satisfies predicate P, that is $P(\boldsymbol{G})$ is true. It means that the system of Eq. (9.14) has a solution in the group $\boldsymbol{G}$, that is there exist $g_1, g_2, \ldots, g_{n_i}$, $i \in I$, $n_i \in N$, such that identities $w_i(g_1, g_2, \ldots, g_{n_i}) = e$ take place for all $i \in I$, and $n_i \in N$ from (9.14).

Let $A = \{g_1, g_2, \ldots, g_{n_i} | i \in I, n_i \epsilon N\}$ be a domain of solutions of the system of Eq. (9.14), that is $A = \{g_1, g_2, \ldots, g_{n_i} | i \in I, n_i \epsilon N, \vDash \bigwedge_{I \in i} w_i(g_1, g_2, \ldots, g_{n_i}) = e\}$. Let's consider a subgroup $\langle A \rangle$ of a group $\boldsymbol{G}$, which is generated by the domain A of solutions of the system of Eq. (9.14). Now let $f{:}\langle A \rangle \rightarrow \boldsymbol{H}$, where $\boldsymbol{H}$ is some group. As f is a homomorphism, then $f(w_i(g_1, g_2, \ldots, g_{n_i})) = w_i(f(g_1), f(g_2), \ldots, f(g_{n_i})) = f(e) = e$, and in view of this, any epimorphic image $\varphi(\boldsymbol{G})$ of a group $\boldsymbol{G}$ satisfies predicate P, as $\varphi(\langle A \rangle) \vDash \bigwedge_{I \in i} w_i (\varphi(g_1), \varphi(g_2), \ldots, \varphi(g_{n_i})) = \varphi(e) = e$. Now let's consider condition (5) and prove the following theorem.

Theorem 9.10 *Next conditions for a group* **G** *are equivalent:*

(1) *if every finite subsystem of the system of* Eq. (9.14) *over a group* above **G** *has a solution in* **G**, *then hole the system of* Eq. (9.14) *has a solution in* **G**,
(2) *a group* **G** *is P-pure injective, where* $P \equiv \bigwedge_{i \in I} w_i(x_1, x_2, \ldots, x_{n_i}) = e$.

Proof We carry out the proof according to the following scheme. Let's prove that (2) implies (1). Let there is a P-pure short exact sequence

$$\begin{array}{c} 1 \rightarrow \boldsymbol{H} \overset{\varepsilon}{\rightarrow} \boldsymbol{K} \rightarrow \boldsymbol{K}/\boldsymbol{H} \rightarrow 1 \\ \varphi \downarrow \\ \boldsymbol{G} \end{array} \tag{9.15}$$

and a homomorphism $\varphi{:}H \rightarrow G$. Let's prove that there exists a homomorphism $\psi{:}K \rightarrow G$ such that $\psi\varepsilon = \varphi$. Let $\{h_j | j \in J\}$ be a system of generators of a group $\boldsymbol{K}$ by $\boldsymbol{H}$ and $\bigwedge_{i \in I} w_i(h_1, h_2, \ldots, h_{n_i}) = e, n_i \in N$, where $i \in I$, be all relations between elements $h_j, j \in J$, and the elements of the group $\boldsymbol{K}$. Let's consider the system of equations

$$\bigwedge\nolimits_{i \in I} w_i(x_1, x_2, \ldots, x_{n_i}) = e \tag{9.16}$$

where $i \in I$, $n_i \in N$. Every finite subsystem of this system of equations contains only finite number of variables $x_{j_1}, x_{j_2}, \ldots, x_{j_k}$. As $\boldsymbol{H}$ is a P-pure subgroup of the group $\boldsymbol{K}$ and $\boldsymbol{H}' = \langle H', h_{j_1}, h_{j_2}, \ldots, x_h \rangle$ is a subgroup of $\boldsymbol{K}$, then $\boldsymbol{H}$ is a P-pure subgroup of the group $\boldsymbol{H}' = \langle H', h_{j_1}, h_{j_2}, \ldots, x_h \rangle$ and so, $\boldsymbol{H}$ is a retract of $\boldsymbol{H}' = \langle H', h_{j_1}, h_{j_2}, \ldots, x_h \rangle$, as by Theorem 9.7 from the condition (1) of the Theorem 9.7 follows condition (2) of the Theorem 9.7. The latter means that there exists an endomorphism σ of the group $\boldsymbol{H}'$ such that $\sigma^2 = \sigma$ and the image σ in $\boldsymbol{H}'$ equals $\boldsymbol{H}$, that is $(\exists \sigma \epsilon End\ H')((\sigma^2 = \sigma) \wedge (Im\sigma = H))$, and the images of components $h_{j_1}, h_{j_2}, \ldots, x_h$ in H under the homomorphism φ give the decision in the group $\boldsymbol{G}$.

So, the system

$$\bigwedge_{i \in I} w_i(x_1, x_2, \ldots, x_{n_i}) = e$$

satisfies the assumptions of (1) of Theorem 9.10, from this we find out that in the group $\boldsymbol{G}$ there exists the decision $x_j = g_j$ of the hole system

$$\bigwedge_{i \in I} w_i(x_1, x_2, \ldots, x_{n_i}) = e.$$

Here $j = 1, \ldots, n_i, i \in I, n_i \in N$.

Further we need the following definition.

Definition 9.11 Let we have the system of equations of the signature $\Omega = \langle ^\circ, \Box^{-1}, e \rangle$:

$$\bigwedge_{i \in I} w_i(x_1, x_2, \ldots, x_{n_i}) = e, \tag{9.17}$$

where $n_i \in N$. The group $F_P = \langle x_1, x_2, \ldots, x_{n_i} | w_i(x_1, x_2, \ldots, x_{n_i}), n_i \in N, i \in I \rangle$ is called a group of solutions of the system of Eq. (9.17).

From the above arguments we obtain that any epimorphic image of the group of solutions F_P of the group of Eq. (9.17) satisfies the system of Eq. (9.17).

With the help of systems of equations, it is possible to write down various relationships between the factors that determine the system S. The first question that arises in this connection is the following one.

How to expound the system of equations:

$$\bigwedge_{i \in I} w_i(x_1, x_2, \ldots, x_{n_i}) = e, i \in I, n_i \in N \tag{9.18}$$

9.6 Interpretation of Systems of Equations Over Groups of Factors that Describe a Smart System

Let $\boldsymbol{G_S}$ be a group of factors that describes the system S, and P be a predicate which is given by a system of equation of the signature $\Omega = \langle ^\circ, \Box^{-1}, e \rangle$:

$$\bigwedge_{i \in I} w_i(x_1, x_2, \ldots, x_{n_i}) = e, \tag{9.19}$$

where $n_i \in N$, and $F_P = \langle x_1, x_2, \ldots, x_{n_i} | w_i(x_1, x_2, \ldots, x_{n_i}), n_i \in N, i \in I \rangle$ be a group of decisions of the system of Eq. (9.19). Let the system (9.19) has a decision in the group $\boldsymbol{G_S}$.

The combination of factors $w_i(x_1, x_2, \ldots, x_{n_i})$ acts on the system S in exactly the same way as the neutral factor e, that is, the word, and in more familiar terminology, the combination of factors, $w_i(x_1, x_2, \ldots, x_{n_i})$ balances the factors $x_1, x_2, \ldots, x_{n_i}$, so the relations

$$\bigwedge\nolimits_{i \in I} w_i(x_1, x_2, \ldots, x_{n_i}) = e$$

we shall call the equilibrium relations of the model $\boldsymbol{G_S}$ of the system S.

The search for systems of equations having a solution in the group $\boldsymbol{G_S}$, can help in the management of the system S in terms of neutralizing the effect of undesirable factors on the system. The following assertions are true.

Theorem 9.12 *Let $\boldsymbol{G_S}$ be a group of factors that describing the system S, and P be a predicate given by a system of equations of the signature* $\Omega = \langle \circ, ^{-1}, e \rangle$:

$$\bigwedge\nolimits_{i \in I} w_i(x_1, x_2, \ldots, x_{n_i}) = e, \tag{9.20}$$

where $n_i \in N$, *and* $F_P = \langle x_1, x_2, \ldots, x_{n_i} | w_i(x_1, x_2, \ldots, x_{n_i}), n_i \in N, i \in I \rangle$ *be a group of solutions of the system of Eq. (9.20). Let the system (9.1) has a solution in the group $\boldsymbol{G_S}$. Then:*

- *any factor group of the group of solutions of the system of equations determines the equilibrium relations of the model $\boldsymbol{G_S}$,*
- *a model $\boldsymbol{H_S}$ of the system S satisfies the equilibrium relations*

$$\bigwedge\nolimits_{i \in I} w_i(x_1, x_2, \ldots, x_{n_i}) = e \tag{9.21}$$

if and only if $\boldsymbol{H_S}$ contains a subgroup which is isomorphic to a factor-group of the group of decisions of the system of equation

$$\bigwedge\nolimits_{i \in I} w_i(x_1, x_2, \ldots, x_{n_i}) = e.$$

Let us note that when modeling a smart system S using a group of factors $\boldsymbol{G_S}$, defining the system S, we usually use a finite number of factors. Therefore the predicate P, if it is given by a system of equations, in this case can be given by a finite system of equations equivalent to the original one, that is, the system

$$\bigwedge\nolimits_{i \in I} w_i(x_1, x_2, \ldots, x_{n_i}) = e \tag{9.22}$$

is equivalent in this case to the system of equations

$$\bigwedge\nolimits_{i=1}^{k} w_i(x_1, x_2, \ldots, x_{n_i}) = e \tag{9.23}$$

where $k \in N$.

Therefore, the group of solutions of the system of Eq. (9.20) will be finite in this case.

Examples of the description of the functioning of systems with the help of systems of equations over groups of factors

(1) The system S achieves a state of equilibrium in the case when the factors determining it commute. In this case a predicate P is given by a system of equations

$$\bigwedge_{i_1, i_2 \in I} x_{i_1} x_{i_2} x_{i_1}^{-1} x_{i_2}^{-1} = e, \tag{9.24}$$

and a group of solutions of the system of Eq. (9.20) is given in the following way:

$$F_P = \left\langle x_1, x_2, \ldots, x_{n_i} | \bigwedge_{i_1, i_2 \in I} x_{i_1} x_{i_2} x_{i_1}^{-1} x_{i_2}^{-1}, i_1, i_2 \in I \right\rangle \tag{9.25}$$

For example, in this way, it is possible to simulate the component of the model of balanced study of the subjects of the training course, the equable of the use of testing and the oral survey in the control of knowledge.

9.7 *P*-Topology

Further we need the following well-known concepts of topology.

Definition 9.11 [5] Let A be a non-empty set and $\mathcal{C}(A)$ be a closure system of the set A, that is $\mathcal{C}(A)$ is a system of subsets of the set A, closed with respect to intersections: $\bigcap B \in \mathcal{C}(A)$, if $B \subseteq \mathcal{C}(A)$.

The closure system is a complete lattice with respect to the inclusion.[2]

Definition 9.12 [5] The closure operator on the set A is the mapping of the set of all subsets $\mathcal{P}(A)$ of the set A into itself $J{:}\mathcal{P}(A) \to \mathcal{P}(A)$ such that:

1. If $X \subseteq Y$, then $J(X) \subseteq J(Y)$,
2. $X \subseteq J(X)$,
3. $JJ(X) = J(X)$
   for all $X, Y \subseteq \mathcal{P}(A)$.

In [5] the following theorem is proved.

Theorem 9.13 *Each closure system* $\mathcal{C}(A)$ *of the set A defines a closure operator J by the rule*

$$J(X) = \bigcap \{Y \in \mathcal{C}(A) | Y \supseteq X\}$$

[2]Let us recall that the lattice (formerly known as structure) is a partially ordered set in which every two-element subset has exact upper (sup) and exact lower (inf) faces. Hence the existence of these faces for any non-empty finite subsets follows. The necessary data on lattices is presented in Chap. 1.

Vice versa, every closure operator J on the set A defines a closure system $\mathcal{C}(A) = \{X \subseteq A | J(X) = X\}$. The correspondence defined in such a way between closure systems and operators is one-to-one. We quote, following [5, p. 59], the Galois correspondence as an example of a closure system.

Let A and B be sets and $\Phi \subseteq A \times B$.

For every $X \subseteq A$ let's define a subset $X^* \subseteq B$ by the equality

$$X^* = \{y \in B | (x, y) \in \Phi \text{ for all } x \in X\}.$$

For every subset $Y \subseteq B$ let's define the subset $Y^* \subseteq A$ by the equality

$Y^* = \{x \in A | (x, y) \in \Phi \text{ for all } y \in Y\}.$

Thus the mappings

$$X \mapsto X^* \text{ and } Y \mapsto Y^* \tag{9.26}$$

of the set $\mathcal{B}(A)$ of all subsets of the set A into the set $\mathcal{B}(B)$ of all subsets of the set B, and $Y \mapsto Y^*$ of the set $\mathcal{B}(B)$ of all subsets of the set B into the set $\mathcal{B}(A)$ of all subsets of the set A, which satisfy the conditions:

if $X_1 \subseteq X_2$, then $X_1^* \supseteq X_2^*$

if $Y_1 \subseteq Y_2$, then $Y_1^* \supseteq Y_2^*$

$$X \subseteq X^{**}, Y \subseteq Y^{**}, \tag{9.27}$$

$$X^{***} = X^*, Y^{***} = Y^*$$

are constructed.

The mappings (9.26) that satisfy conditions (9.27) are called the Galois correspondence.

By analogy with Definitions 9.11 and 9.12, let us formulate the definitions of a system of P-closures and the operator P-closures on a group for the predicate P defined on the class of all groups closed with respect to taking subgroups and factor groups.

Definition 9.14 [5] A closure operator on a set A is said to be algebraic if for any $X \subseteq A$ and $a \in A$, $a \in J(X)$ implies $a \in J(X_j)$ for some finite subset X_j of the set X.

Definition 9.15 Let P be a predicate given on the class of all groups which is closed under taking subgroups and factor-groups and $\boldsymbol{A} = \langle A, ^\circ, \Box^{-1} \rangle$ be a group. Let's denote by $CP(\boldsymbol{A})$ the system of subgroups of a group $\boldsymbol{A}$, which are closed under intersections and satisfy the predicate P: $\cap \boldsymbol{B} \in CP(\boldsymbol{A})$, if $\boldsymbol{B} \preccurlyeq CP(\boldsymbol{A})$, where by the symbol $\preccurlyeq$ the relation "to be a subgroup of a group" is designated.

Definition 9.16 Let P be a predicate given on the class of all groups and closed under taking subgroups and factor-groups. Let's call the mapping $JP{:}\mathcal{P}(\boldsymbol{A}) \to \mathcal{P}(\boldsymbol{A})$ of the set $\mathcal{P}(\boldsymbol{A})$ of all subgroups of a group $\boldsymbol{A}$ into itself the P-closure operator on the group $\boldsymbol{A} = \langle A, ^\circ, \Box^{-1} \rangle$ such that

1. If $\boldsymbol{X} \preccurlyeq \boldsymbol{Y}$, then $JP(\boldsymbol{X}) \preccurlyeq JP(\boldsymbol{Y})$,
2. $\boldsymbol{X} \preccurlyeq JP(\boldsymbol{X})$,
3. $JP(JP(\boldsymbol{X})) = JP(\boldsymbol{X})$ for all $\boldsymbol{X}, \boldsymbol{Y} \preccurlyeq \boldsymbol{A}$,
   where by the symbol $\preccurlyeq$ the relation "to be a subgroup of" is denoted.

By analogy with [5], we can prove a theorem on a one-to-one correspondence between closure systems $CP(\boldsymbol{A})$ by the predicate P and P-closure operators JP, defined by the rule:

$$JP(\boldsymbol{X}) = \bigcap \{\boldsymbol{Y} \in CP(A) | Y \succcurlyeq X\}$$

9.8 Pro-*P*-Algebraic Systems

9.8.1 Inverse and Direct Spectra of Algebraic Systems and Their Limits

We shall need the definition of direct and inverse spectra of algebraic systems and their limits for the sequel.

Inverse spectra.

We follow [1] in presenting this material. Let us note that the notion of an inverse limit is dual with respect to the notion of a direct limit.

Definition 9.17 Let's consider the class of all algebras of the signature $\Omega = \{f_\alpha^{n_\alpha} | \alpha \in \Gamma\}$. Let $\{\bar{A}_i = A_i | \{f_\alpha^{n_\alpha} | \alpha \in \Gamma\} | i \in I\}$ be a set of algebraic systems of the signature $\Omega = \{f_\alpha^{n_\alpha} | \alpha \in \Gamma\}$, partially ordered set of indices I is directed, that is for every $i, j \in I$ there exists $k \in I$ such that $i \leq k, j \leq k$. Let for every pair of indices $i, j \in I$ such that $i \leq j$, homomorphism

$\pi_i^j{:}\bar{A}_j = \langle A_j|\{f_\alpha^{n_\alpha}|\alpha \in \Gamma\}\rangle_j \rightarrow \langle A_i|\{f_\alpha^{n_\alpha}|\alpha \in \Gamma\}\rangle$ is given, and the following conditions hold:

(1) π_i^i is the identity map of an algebraic system $\langle A_i|\{f_\alpha^{n_\alpha}|\alpha \in \Gamma\}\rangle$ for every $i \in I$,
(2) for every $i,j,k \in I$, such $i \leq j \leq k$ we have the equality $\pi_i^j \cdot \pi_j^k = \pi_i^k$.

Then the system $\{\{\bar{A}_i = \langle A_i|\{f_\alpha^{n_\alpha}|\alpha \in \Gamma\}\rangle\}||i \in I, \pi_i^j\}$ is called an inverse spectrum.

The inverse (or projective) limit or simply the limit of the inverse spectrum $\{\{\bar{A}_i = \langle A_i|\{f_\alpha^{n_\alpha}|\alpha \in \Gamma\}\rangle\}||i \in I, \pi_i^j\}$ is called an algebraic system $\bar{A}^* = \lim\limits_{\leftarrow} \bar{A}_i$ of the direct product $A = \prod_{i \in I} A_i$, which consists of all vectors $a = (\ldots, a_i, \ldots, a, \ldots)$, such that $\pi_i^j a_j = a_i, i \leq j$.

We also need the following well-known facts concerning projective limits:

1. There exist homomorphisms $\pi_i{:}\bar{A}^* \rightarrow A_i, i \in I$, such that all diagrams where $i \leq j$

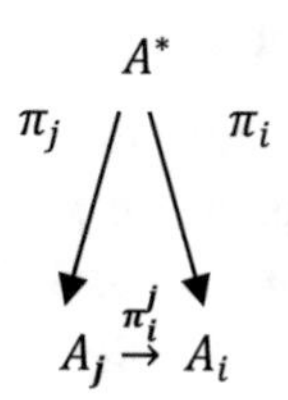

are commutative.

2. The inverse limit $\bar{A}^*$ of an inverse spectrum $\{\bar{A}_i|i \in I, \pi_i^j\}$ poses the following property: if A is an algebraic system and $\sigma_i{:}A \rightarrow \bar{A}_i$ are homomorphisms for which the diagrams

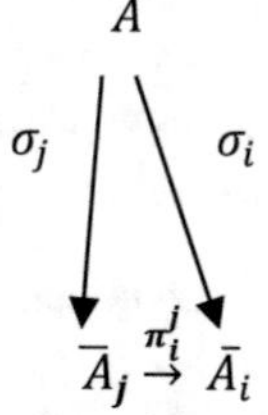

are commutative, then there exists the only homomorphism $\sigma{:}A \rightarrow \bar{A}^*$, such that all the diagrams

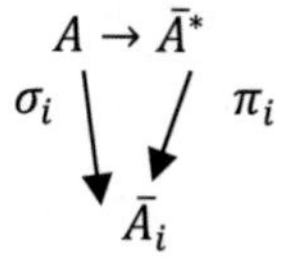

$A \rightarrow \bar{A}^*$
$\sigma_i \searrow \quad \swarrow \pi_i$
$\bar{A}_i$

where $\pi_i:g \to g_i$ is a canonical homomorphism that is a restriction of the i-th coordinate projection of an algebraic system $\prod_{i \in I} \bar{A}_i$ on $\bar{A}^*$. This property defines an inverse limit up to isomorphism, [4, 6].

Definition 9.18 Let's consider the class of all algebras of the signature $\Omega = \{f_\alpha^{n_\alpha} | \alpha \in \Gamma\}$. Let P be a predicate given on the class of all algebras of the signature $\Omega = \{f_\alpha^{n_\alpha} | \alpha \in \Gamma\}$, which is closed with respect to subalgebras and factor-algebras. An algebra $\bar{A} = \langle A | \{f_\alpha^{n_\alpha} | \alpha \in \Gamma\} \rangle$ of the signature $\Omega = \{f_\alpha^{n_\alpha} | \alpha \in \Gamma\}$ is called a P-finite if $P(G)$ is true and an algebra $\bar{A} = \langle A | \{f_\alpha^{n_\alpha} | \alpha \in \Gamma\} \rangle$ is finite.

Definition 9.19 Let's consider the class of all algebras of the signature $\Omega = \{f_\alpha^{n_\alpha} | \alpha \in \Gamma\}$. Let P be a predicate given on the class of all algebras of the signature $\Omega = \{f_\alpha^{n_\alpha} | \alpha \in \Gamma\}$, which is closed with respect to and factor-algebras. An algebra $\bar{A} = \langle A | \{f_\alpha^{n_\alpha} | \alpha \in \Gamma\} \rangle$ of the signature $\Omega = \{f_\alpha^{n_\alpha} | \alpha \in \Gamma\}$ is called P-profinite, (or pro-P-algebra of the signature $\Omega = \{f_\alpha^{n_\alpha} | \alpha \in \Gamma\}$), if it is isomorphic to an inverse limit of P-finite algebras of the signature $\Omega = \{f_\alpha^{n_\alpha} | \alpha \in \Gamma\}$.

Definition 9.20 Let's consider the class of all algebras of a signature $\Omega = \{f_\alpha^{n_\alpha} | \alpha \in \Gamma\}$. Let P be a predicate given on a class of all algebras of a signature $\Omega = \{f_\alpha^{n_\alpha} | \alpha \in \Gamma\}$, closed with respect to subalgebras and factor-algebras. An algebra $\bar{A} = \langle A | \{f_\alpha^{n_\alpha} | \alpha \in \Gamma\} \rangle$ of a signature $\Omega = \{f_\alpha^{n_\alpha} | \alpha \in \Gamma\}$ is called a P-algebraically compact if it is retract of any algebra $\boldsymbol{G}$, which contains it as a P-pure subalgebra, such that A^2 defines congruence (kernel equivalence), [7], on $\boldsymbol{G}$.

Theorem 9.21 is an analogues of theorem from [1].

Theorem 9.21 *Let's consider the class of all algebras of a signature* $\Omega = \{f_\alpha^{n_\alpha} | \alpha \in \Gamma\}$. *Let* P *be a predicate given on a class of all algebras of a signature* $\Omega = \{f_\alpha^{n_\alpha} | \alpha \in \Gamma\}$, *closed with respect to subalgebras and factor-algebras. The following conditions for an algebra* $\bar{A} = \langle A | \{f_\alpha^{n_\alpha} | \alpha \in \Gamma\} \rangle$ *of a signature* $\Omega = \{f_\alpha^{n_\alpha} | \alpha \in \Gamma\}$ *are equivalent:*

(1) *An algebra* $\bar{A}$ *is P-algebraically compact;*
(2) *An algebra* $\bar{A}$ *is P-pure injective;*
(3) *An algebra* $\bar{A}$ *is a retract of a P-profinite subalgebra of a signature* $\Omega = \{f_\alpha^{n_\alpha} | \alpha \in \Gamma\}$ *(inverse limit of pro- P-algebras of a signature* $\Omega = \{f_\alpha^{n_\alpha} | \alpha \in \Gamma\}$*).*

The proof of this theorem repeats the proof of Theorem 9.8 and Definition 9.9 enables us to consider predicates defined by systems of equations on the class of all algebras of signature $\Omega = \{f_{\alpha}^{n_{\alpha}} | \alpha \in \Gamma\}$.

References

1. Fuchs, L.: Infinite Abelian groups, vol. 1. Mir, Moscow (1974) (in Russian)
2. Reid, C.D.: Finiteness properties of profinite groups. A thesis submitted by the University of London for the degree of Doctor of Philosophy, Queen Mary, University of London, 2010, February 10, LNCS Homepage. http://www.maths.qmul.ac.uk/
3. Ershov, Yu.L: Profinite groups. Algebra Logic **19**(5), 552–565 (1980)
4. Artamonov, V.A., Salii, V.N., Skornyakov, L.A., Shevrin, L.N., Shul'geifer, E.G.: General Algebra, vol. 2, p. 188. Science. Main Edition of the Physical and Mathematical Literature, Moscow (1991) (in Russian)
5. Cohn, P.M.: Universal algebra. Harper & Row Publishers, New York-London, N.Y. (1965)
6. Bucur, I., Deleanu, A.: Introduction to the Theory of Categories and Functors, Pure and Applied Mathematics, vol. 19. Interscience Publishers (1968)
7. Malt'sev, A.I.: Algebraic systems. Nauka, Moscow (1970) (in Russian)

Chapter 10
P-Sustainability of a System. Algebraic Formalization of Sustainability Concept. Sustainability of Ranking Systems in Education

Abstract The question of the reliability of the obtained results is of great value for any theory. This is especially important when it comes to risk-free application of the theoretical results in practice. The reliability is especially significant for the humanities relating to the development and functioning of human society, such as pedagogy, the general theory of education, e-learning, economics, finance and so on as their distinctive features are the following:

- impossibility of repetition the experiment and frequently to perform the only experiment with sufficient accuracy, since there is always the human factor,
- the difficulty of collecting reliable and comparable statistical data in connection with the lack of standardized procedures. In this chapter we continue to study smart systems, and in particular, the concept of smart—university in the context of theoretical justification of the results based on the algebraic formalization of the smart systems. The practice result of this investigation is the evaluation of sustainability of ranking universities systems.

Keywords Sustainability · Smart education · Ranking systems

10.1 Sustainability: Ranking Systems

In the study of system's functioning across the time and its ability to forecast changes of system's properties the question about system sustainability is rather important. This question is especially important for the Smart System Theory. The concept of sustainability is well—studied in terms of the availability of various quantitative parameters describing the dynamic behavior of the system [1, 2]. There were introduced such concepts as Lyapunov sustainability, Zhukovsky sustainability and so on. We shall consider discrete systems as in previous chapters. Under the sustainability of a discrete system we shall understand its ability to return to the equilibrium position after the end of the action of external factors as in the case of continuous—time systems. To date the classification of such concepts as an equilibrium, as a notion of stationary point and so on there were introduced. The indices characterizing the

N. Serdyukova and V. Serdyukov, *Algebraic Formalization of Smart Systems*,
Smart Innovation, Systems and Technologies 91,
https://doi.org/10.1007/978-3-319-77051-2_10

quality of discrete systems designed to evaluate the dynamic properties of the system, manifested in transient conditions and to determine the accuracy of the system which is characterized by errors in the steady state after the transition were introduced [3]. Dynamic indicators of quality characterize the behavior of the free components of the transition process closed control systems or processes of an autonomous system. Herewith the stable systems only are considered. However convenient integrated indicators which are a synthesis of qualitative and quantitative indicators of the phenomenon under study as such are absent. We propose to use Cayley table of a group G_S of factors determining the system S to characterize the quality of dynamics of the closed associative smart system with feedback S. This makes it possible to regulate the behavior of the smart system S in some cases. The time factor is introduced into the construction of the groups of factors G_S determined the system S for this purpose. The concept of the final sustainability of a system is defined and the link between the final sustainability and Lyapunov sustainability is reviewed. An algorithm to determine the points (intervals) of the loss of a sustainability of a system S and scenarios of functioning of a system S (Algorithm 10.9) is constructed. Examples of a usage of parametric statistic in part of laws of distribution of discrete random variables in an annex to the scenarios of development of the system S are discussed. An algorithm to define and regulate scenarios of system's functioning where a system is defined by a group of factors G_S of order p^2 for a prime number p (Algorithm 10.9) is built. The connection between of the notion of final stability, stationary points and the classical notion of sustainability is discussed.

10.2 Final Sustainability of a System

Let us formulate the notion of a final sustainable system, which in fact is an analogue of the classical notion of a sustainable equilibrium.

Definition 10.1 Let $\boldsymbol{G_S} = \langle G_S, \circ, \square^{-1}, e\rangle$ be a group of factors which represent the system S. Let G_S be finite and $|G_S| = n$. Let $G_{1S}, G_{2S}, \ldots, G_{mS}$ be all pair wise non isomorphic groups of n elements. Groups $\boldsymbol{G_{1S}}, \boldsymbol{G_{2S}}, \ldots, \boldsymbol{G_{mS}}$ are called final states of a system S. A system S is called a final sustainable one if it has only one final state.

Definition 10.2 Groups $\boldsymbol{G_{1S}}, \boldsymbol{G_{2S}}, \ldots, \boldsymbol{G_{mS}}$ are also called scenarios of a development of a system S.

The following statements follow directly from the Definitions 10.1 and 10.2.

Corollary 10.3 *System S, which is represented by a simple groups of factors $\boldsymbol{G_S}$ of the order less than 1000000 is final sustainable. A system S which is represented by a groups of factors $\boldsymbol{G_S}$, which consists of a finite prime number of factors, is final sustainable.*

The next problem (that arises in this context) is about how the time factor can be entered into a design of a group of factors—those that represent the associative closed system with a feedback.

10.3 Time Structure of Algebraic Formalization

Let us introduce the time factor into the construction which represents an associative closed system with a feedback S.

The time factor is represented by a linearly ordered set $\langle T, \leq \rangle$ in [3]. Let us interpret time a little bit differently, namely, by considering an abelian linearly ordered group $\langle \boldsymbol{T} = T, +, -, 0 \leq \rangle$. In this interpretation, we assume the datum point from "the modern period *0*", and, therefore, we can review past time periods with the help of the operation -.

Let $\boldsymbol{G_S} = \langle G_S, \circ, \square^{-1}, e \rangle$ be a group of factors which represent the system S. Let's consider the Cartesian product of sets $\boldsymbol{G_S} \times \boldsymbol{T}$.

We shall use the following notations:

An ordered pair $\langle a, t \rangle$ will be denoted as $a(t)$, that is $a(t) \rightleftharpoons \langle a, t \rangle$.

The connections between factors which represent the system S and the factor time are defined by binary operations $\oplus$ and $*$ on $G_S \times T$ satisfying the following conditions:

(1) $a(x) \oplus a(y) = a(x+y)$, for every $a \in G_S$, for every $x, y \in T$,
(2) $a(x) * b(x) = (a \circ b)(x)$ for every $a, b \in G_S$, for every $x \in T$.

So we have that every $a \in G_S$ defines the homomorphism $\breve{a} : T \to G_S \times T$, such that for any $x \in T$ we have $x \xrightarrow{\breve{a}} a(x)$. For a complete inverse image of $M \subseteq G_S \times T$ under the map $\breve{a}$ we have the inclusion $(\breve{a})^{-1}(M) \subseteq T$. Every $x \in T$ defines a homomorphism $\widehat{x} : G_S \to G_S \times T$, such that for any $a \in G_S$ we have: $a \xrightarrow{\widehat{x}} a(x)$. For a complete inverse image of $M \subseteq G_S \times T$ under the map $\widehat{x}$ the inclusion $(\widehat{x})^{-1}(M) \subseteq G_S$ takes place. Now let $T(t_1; t_2)$ be the minimal subgroup on the relation of inclusion which contains all $x \in T$ such that $t_1 \leq x \leq t_2$. Let $G_S \times T(t_1; t_2) \subseteq M \subseteq G_S \times T$ and $x \in T$. Then we have the inclusion: $(\hat{x}^{-1})(G_S \times T(t_1; t_2)) \subseteq (\hat{x}^{-1})(M) \subseteq G_S$. A right action ρ of a group G_S on $(\widehat{x}^{-1})(M) \subseteq G_S$ is a map $\rho : G_S \times (\widehat{x}^{-1})(M) \to (\widehat{x}^{-1})(M)$, such that $(a, g) \mapsto ag$, $ae = a, ag_1g_2 = (ag_1)g_2$, where $a \in G_S$, g, g_1, $g_2 \in (\widehat{x}^{-1})(M)$. A right action ρ of a group G_S on the set $(\widehat{x}^{-1})(M) \subseteq G_S$ defines the homomorphism $G_S \to Symm((\widehat{x}^{-1})(M))$, for which $a \mapsto ag$, where $a \in G_S, g \in (\widehat{x}^{-1})(M)$, where $Symm((\widehat{x}^{-1})(M))$ is a group of all permutations on the set $((\widehat{x}^{-1})(M)) \subseteq G_S$. The group $Symm((\widehat{x}^{-1})(M))$ is a subgroup of the group $Symm\, G_S$. Let $i : G_S \to Symm\, G_S \cong S_n$, where S_n is a symmetric group of all permutations of degree n. Let's designate an isomorphism $Symm\, G_S \cong S_n$ by $\propto$. Let A_n be an alternating group of permutations of degree n. Let us choose a set M in such a way that the inclusion $i^{-1} \propto^{-1} (A_n) \subseteq M$ takes place. Then $(\widehat{x})^{-1}(i^{-1} \propto^{-1} (A_n)) \subseteq (\widehat{x}^{-1})$ (M) и $(\widehat{x}^{-1})(M)$ is a simple group. The last statement follows from the Diagram 10.1.

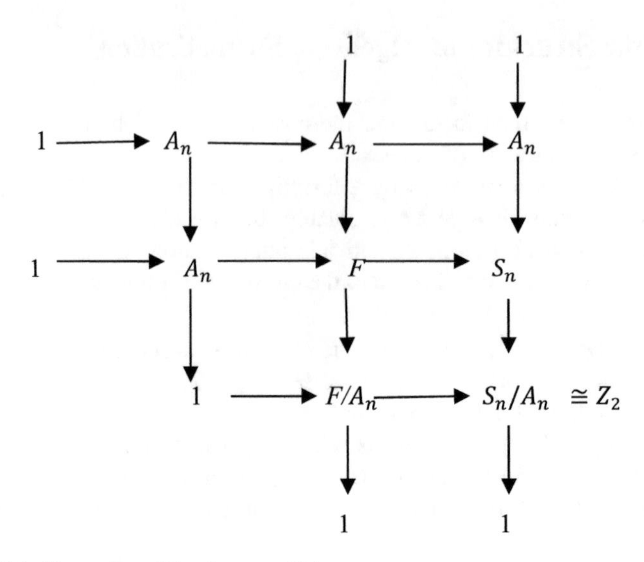

Diagram 10.1 Illustration of the theorem 10.4

We have

$$F/A_n \cong 1 \text{ or } F/A_n \cong Z_2, \text{ so } F \cong A_n \text{ or } F \cong S_n.$$

Hence we obtain the following statement.

Theorem 10.4 *A right action of the group of factors* $\mathbf{G_S}$*, which determines the system S on the set* $(\widehat{x}^{-1})(M)$*, where* $i^{-1} \propto^{-1} (A_n) \subseteq M$*, defines a single scenario of the development of the system S for any instant of time x.*

Let's clear up the sense of this theorem.

If at any point of time t_0 the set of factors acts on the system S which contains in a specific way, namely as $i^{-1} \propto^{-1} (A_n)$, a subgroups of factors isomorphic to A_n, then in the moments of time $t \geq t_0$ the system S is developed by the unique scenario.

Theorem 10.5 *In order to make it possible to adjust the scenario of a development of a system S it requires the existence the instant of time* $x \in T$ *such that for any* $M \subseteq G_S \times T$ *the condition* $i^{-1} \propto^{-1} (A_n) \not\subseteq M$ *is performed.*

Let's write the last condition in another way:

$$(\exists x \in T)(\forall M \subseteq G_S \times T)(\exists y)\left(\left(y \in i^{-1}\alpha^{-1}(A_n) \wedge (y \notin M)\right)\right) \tag{10.1}$$

Definition 10.6 An element $x \in T$ such that the following conditions are true:

(1) $i^{-1} \propto^{-1} (A_n) \subseteq M$
(2) x is a minimal one by the relation $\leq$

is called the point of no return of the system S.

Remark The inclusion $i^{-1} \propto^{-1} (A_n) \subseteq M$ depends on the parameter x.

Definition 10.7 The system S is called sustainable in the period of time $t_1 \le t \le t_2$, if S does not change the scenario of development in this interval.

The following question arises: how to determine points and interval of loss of the sustainability of a system? The answer to this question is given by Sect. 10.4

10.4 The Algorithm of Determination the Scenarios of Development of the System S and Points and Intervals of Loss of the Sustainability of the System S

Let $G_S = \{a_i | i = 1, \ldots, n\}$, $|G_S| = n$, and $G_{1S}, G_{2S}, \ldots, G_{mS}$ be all pair wise non isomorphic groups of n elements, $T_{1S}, T_{2S}, \ldots, T_{mS}$ be Cayley tables respectively for groups $G_{1S}, G_{2S}, \ldots, G_{mS}$. Let the equality

$$a_i \circ a_j = a_{rk} \tag{10.2}$$

where $\in \{1, \ldots, n\}$, is true in the Cayley table T_{kS}.

1. Let we have all indices $a_i(t), a_j(t), a_{rk}(t)$ for $t_1 \le t \le t_2$. We shall assume that every indices from $a_i(t), a_j(t), a_{rk}(t)$ does not change the nature of monotony in the time interval $t_1 \le t \le t_2$. Then there are two of these indices that have the same nature of monotony. If these are a_i and a_j, and a_{rk} has another nature of monotony then in the time interval $t_1 \le t \le t_2$ Cayley table T_{kS} does not work. This means that the algebraic formalization G_{kS} does not work anymore and we should choose possible ways of modeling the development of the system S among those Cayley tables $T_{1S}, T_{2S}, \ldots, T_{mS}$ in which a_i, a_j, a_{rk} have the same nature of monotony. Remembering lemma about compressing segments and appropriately diminishing the interval $t_1 \le t \le t_2$, we get the point of loss sustainability or the interval of loss of sustainability that depends of the quality of statistic data of the system S. Let the scenario of development G_{kS} took place in the time interval $t_0 \le t \le t_1$. Then the ratio of the lengths $q = \frac{|[t_1;t_2]|}{|[t_0;t_2]|}$ of the time intervals shows the probability of default the scenario of development G_{kS} of the system S. Let's show that the ratio of the lengths of the time intervals defines a probability measure. In fact let $X = \{G_{1S}, \ldots, G_{mS}\}$ be the set of all scenarios of development of a system S, which is described by a finite set of factors. Then the set X is finite. Let $P(X)$ be a set of all subsets of the set X. Then $|X| = 2^m$. The algebra $\langle P(X), \cup, \cap, ' \rangle$ is a finite σ—algebra.
2. In accordance with step 1 we choose all of the scenarios of the development $\{G_{is} | 1 \le i \le m, i \in I_k\}$, for the system S such that for each of which the probability of its default in the time interval $t_1 \le t \le t_2$ is more or equal q.

3. Repeat step 1 for the scenarios of developments for the system S from the set $\{G_{1S},\ldots,G_{mS}\}\backslash\{G_{is}|1\leq i\leq m, i\in I_k\}$.
4. Since the number of possible scenarios of the development of the system S is finite than on some finite step j the process terminates.
5. On the penultimate step $j-1$ we get the scenario of the development $G_{k_{j-1}S}$ for the system S in the time interval $t_1\leq t\leq t_2$.

The nature of monitoring indices of factors which determine the system's S behavior or functioning is discrete, so in further to predict the behavior of the system S to determine the scenario in accordance to which it will develop we need the most well known laws of distribution of random values.

Remark It is easy to see that this algorithm defines points of no return of the system S.

Now we shall construct an algorithm to define and regulate scenarios of functioning of the system S by identifying the points of no return of a system S modeled by a group of factors G_S of an order p^2, where p is a prime number.

10.4.1 The Algorithm of a Determination and Regulation the Scenarios of a Functioning of a System S with a Group of Factors G_S of an Order p^2, Where p Is a Prime Number

To determine and regulate an algorithm of a determination and regulation the scenarios of a functioning of a system S and the points of no return of a system and the scenario in accordance to which the system operates at certain time intervals and some quantitative estimates we shall use parametric statistics in terms of laws of distribution of discrete random variables. Namely we shall use parametric statistic in terms of laws of distribution of discrete random variables in the annex to the scenarios of a development of a system S.

Binomial distribution to perform the system S the development scenario G_{kS} in n trials with the probability of a success equals p

The value of a random variable ξ is equal to the number of successful realizations of model development $\mathbf{G_S}$ by the scenario development G_{kS} in n trials with the probability of realization of the scenario development equals $\boldsymbol{p}$ [4]. Let us assume that $\boldsymbol{p}=1-q$ than we obtain the binomial distribution for the model $\mathbf{G_S}$ to perform the scenario G_{kS} in n trials with a probability of a success equals $\boldsymbol{p}$:

ξ	0	1	…	l	…	n
Probability	$(1-\boldsymbol{p})^n$	$n\boldsymbol{p}(1-\boldsymbol{p})^{n-1}$	…	$C_n^l\boldsymbol{p}^l(1-\boldsymbol{p})^{n-l}$	…	$\boldsymbol{p}^n$

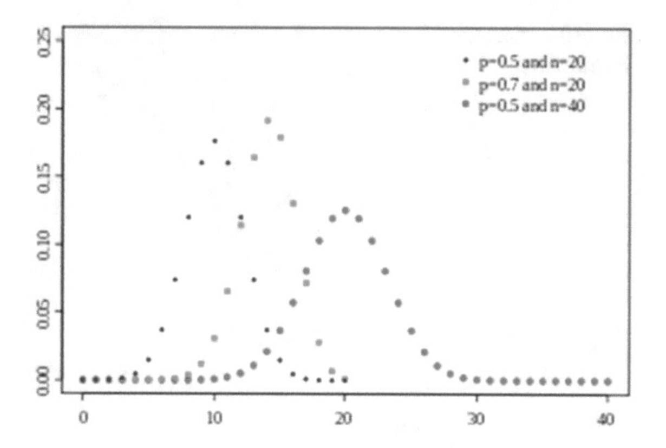

Fig. 10.1 Binomial distribution

Let us explain now the sense of the binomial distribution for modeling the development of a system by a finite number of scenarios and the possibility of its use to simulate operation of a system S. The test of a realization of a scenario G_{kS} is carried out by indicators $D_{ks} = \{d_{\propto ks} | \propto \in \Lambda_{ks}\}$ which are characterizing this scenario allowing one to distinguish between scenarios of the system development S. The binomial distribution characterizes the functioning of a system S with a group of determining factors which consists of $\boldsymbol{p}^2$ elements where $\boldsymbol{p}$ is a prime number, because there exist exactly two non isomorphic groups of the order $\boldsymbol{p}^2$ and each of them are abelian. The mathematical expectation $\mu\xi$ of a random value ξ, distributed by the binomial distribution which is equal to the number of successful realizations of the scenario of functioning G_{kS} of a system S, with the probability $\boldsymbol{p}$ of successful realization the scenario of functioning G_{kS} is equal to $\mu\xi = n\boldsymbol{p}$, the dispersion $\sigma^2(\xi)$ of ξ is equal $\sigma^2(\xi) = n\boldsymbol{p}(1-\boldsymbol{p})$.

Chebyshev's inequality applied to the binomial distribution gives the following [4]

Corollary 10. 8 *For each* $\varepsilon > 0$ *we have the inequality:*
The probability

$$P(|\xi - n\boldsymbol{p}| > \varepsilon) \leq \frac{n\boldsymbol{p}(1-\boldsymbol{p})}{\varepsilon^2}$$

Hence, according to three sigma rule, we obtain:
The probability

$$P\left(n\boldsymbol{p} - 3\sqrt{n\boldsymbol{p}(1-\boldsymbol{p})} \leq \xi \leq n\boldsymbol{p} + 3\sqrt{n\boldsymbol{p}(1-\boldsymbol{p})}\right)$$
$$= 1 - P\left(|\xi - n\boldsymbol{p}| > 3\sqrt{n\boldsymbol{p}(1-\boldsymbol{p})}\right) \geq 8/9 \approx 0.9$$

Let us cite the curve of the binomial distribution (Fig. 10.1):

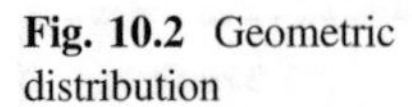
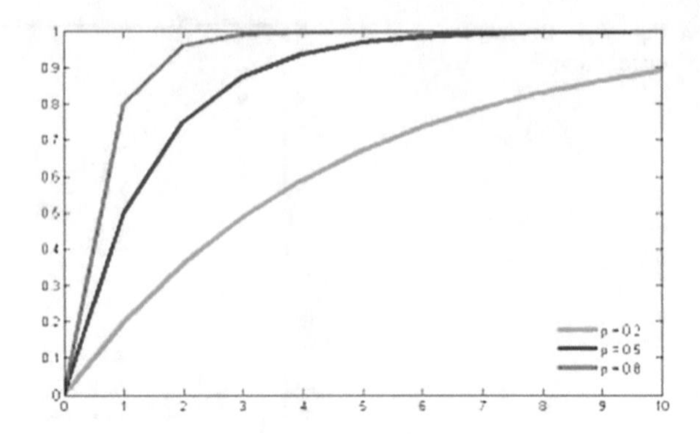

Fig. 10.2 Geometric distribution

Geometric distribution to perform the system *S* the development scenario G_{kS} in *n* trials with the probability of a success equals *p*

The value of the random variable τ is equal to the number of the first successful trial in the Bernoulli scheme with the probability of success equals $\boldsymbol{p}$ including successful ones [4]

τ	1	2	…	l	…
Probability	$\boldsymbol{p}$	$\boldsymbol{p}(1-\boldsymbol{p})$	…	$\boldsymbol{p}(1-\boldsymbol{p})^{l-1}$	…

Let us explain now the sense of the geometric distribution for modeling the development of a system by a finite number of scenarios and the possibility of its use to simulate operation of a system *S*. The test of a realization of a scenario G_{kS} is carried out by indicators $D_{ks} = \{d_{\propto ks} | \propto \in \Lambda_{ks}\}$ which are characterizing this scenario allowing one to distinguish between scenarios of the system development *S*. The geometrical distribution characterizes the functioning of a system *S* with a group of determining factors which consists of $\boldsymbol{p}^2$ elements where $\boldsymbol{p}$ is a prime number, because there exist exactly two non isomorphic groups of the order $\boldsymbol{p}^2$ and each of them are abelian ones. The geometric distribution shows the beginning of the realization of the scenario G_{kS} during the functioning of the system *S*.

Let us cite the curve of the geometric distribution (Fig. 10.2):

The Poisson distribution for the implementation the scenario of a development G_{kS} of a system *S* with parameter λ with the probability of a realization of the scenario of a development of a system *S* equals to *p*

ξ	0	1	…	l	…
Probability	$e^{-\lambda}$	$\lambda e^{-\lambda}$	…	$\frac{\lambda^l}{l!} e^{-\lambda}$	…

The value of the random variable ξ is equal to the number of events having been occurred in a fixed time provided that these events occur with a fixed average intensity λ and independently from each other. Poisson's law is a limit to the binomial distribution if the number of tests *n* tends to infinity, the probability $\boldsymbol{p}$ tends

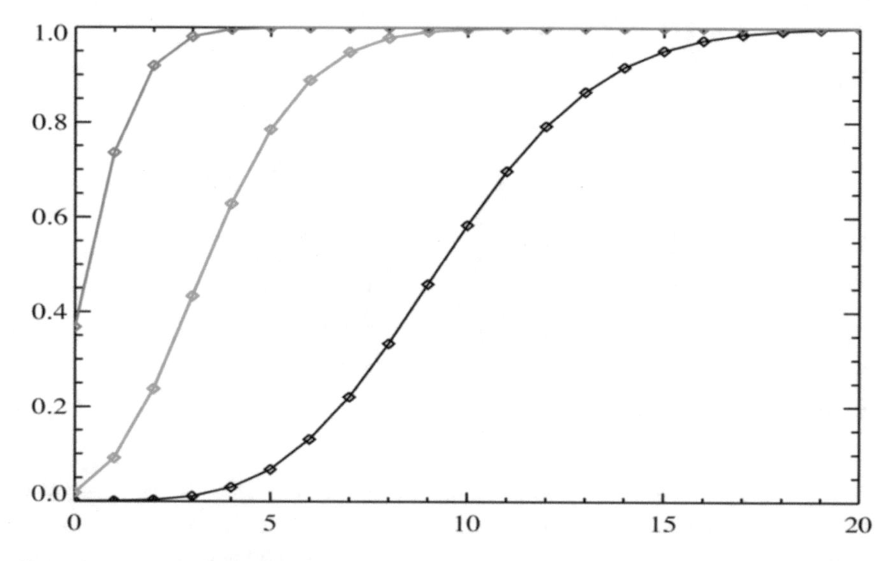

Fig. 10.3 Poisson distribution, λ = 1, 4, 10

to zero in such a way that their product $np = \lambda$. Let us explain now the sense of Poisson distribution for modeling the development of a system by a finite number of scenarios and the possibility of its use to simulate operation of a system S. The test of a realization of a scenario G_{kS} is carried out by indicators $D_{ks} = \{d_{\propto ks} | \propto \in \Lambda_{ks}\}$ which are characterizing this scenario allowing one to distinguish between scenarios of the system development S. The monitoring of indicators D_{ks} allows one to determine, using the Poisson theorem, the probability of a development of a system by a scenario of a development of a system G_{kS}, in case when the probability of the realization of the scenario G_{kS} tends to zero: $\boldsymbol{p} \to 0$, and the number of experiments (the number of observations) tends to infinity: $n \to \infty$ in such a way that $np = \lambda$. The Poisson distribution can be used to identified and prevent unwanted scenarios of the development of a system, that is any exceptions or violations of planned development options a system S.

Let us cite the curve of Poisson distribution (Fig. 10.3):

Hypergeometric distribution for the implementation the scenario of a development G_{kS} of a system S with parameters a, b, n with the probability of a realization of the scenario of a development of a system S equals to p

ξ	0	1	…	m	…	n
Probability	0	$(aC_b^{n-1})/C_{a+b}^n$	…	$(C_a^m C_b^{n-m})/C_{a+b}^n$	…	C_a^n/C_{a+b}^n

$$C_{a+b}^n = \frac{(a+b)!}{n!(a+b-n)!}$$

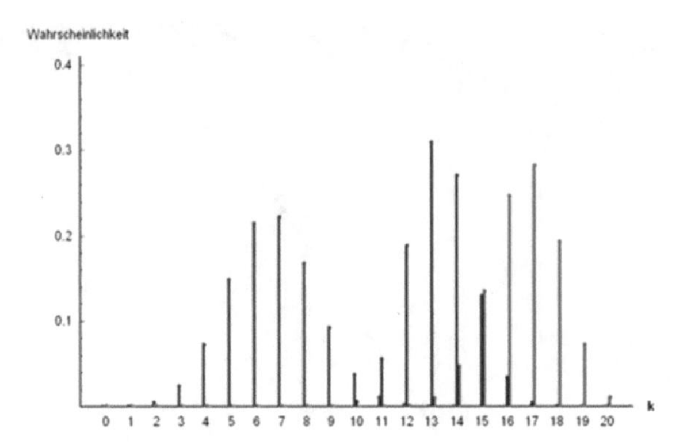

Fig. 10.4 The hypergeometric distribution, λ = 1, 4, 10

The hypergeometric distribution simulates the number of successful samples from the finite population and indicates the probability of the choice from $n = a + b$ elements which are scenarios of functioning of a system S and among them there are a scenarios of the first sort and b scenarios of the second sort. Let us explain now the sense of hypergeometric distribution for modeling the development of a system by a finite number of scenarios and the possibility of its use to simulate operation of a system S. The test of a realization of a scenario G_{kS} is carried out by indicators $D_{ks} = \{d_{\propto ks}| \in \Lambda_{ks}\}$ which are characterizing this scenario allowing one to distinguish between scenarios of the system development S. The hypergeometric distribution can be used to determine the probability of a realization a times a script of the first sort G_{1S} and b times a script of the second sort G_{2S} during the functioning the system S, described by the group of factors of the order p^2, where p is a prime number because there exist exactly two non isomorphic groups of the order p^2 and each of them are abelian ones.

Let us cite the curve of the hypergeometric distribution (Fig. 10.4):

We now construct an alternative Algorithm 10.9 for determining system S functioning scenarios.

The Algorithm 10.9 for Determining System S Functioning Scenarios

1. Let $G_S = \{a_i|i = 1,\ldots,n\}$, $|G_S| = n$, и $G_{1S}, G_{2S},\ldots, G_{mS}$—all pairwise nonisomorphic groups of n elements, $T_{1S}, T_{2S},\ldots, T_{mS}$ be Cayley tables for groups $G_{1S}, G_{2S},\ldots, G_{mS}$ respectively.

   Let's consider all equations of the form

$$a_i \cdot a_j = a_k \tag{10.1}$$

   where a_i and a_j are of the same nature of monotonicity for the time interval $t_0 \leq t \leq t_1$.

   If a_k is of a different character of monotonicity in this time interval, than a group G_{1S} such that in its Cayley table $a_i \cdot a_j = a_k$, does not model a system S in the time interval $t_0 \le t \le t_1$. In order to understand which of the groups of an order n simulates the system S one should select such groups from all groups of an order n those in which the equation

$$a_i \cdot a_j = a_r,$$

takes place and a_i, a_j and a_r are of the same character of monotonicity in the time interval $t_0 \le t \le t_1$.
Let it be groups G_{2S}, G_{3S}, …, G_{mS}.

2. Then in the same way as in the step 1 one should check all the rest of the ratio of the Cayley table of the group G_{2S}.

3. As the group G_S is finite the process terminates on the step with the number not exceeding m.

The following well known facts [5, 6], from the group theory can be used to simulate the number of system's scenarios of functioning:

(1) Let p, q be prime numbers, $p < q$. If p does not divide $q - 1$, then in this case there exists unique up to isomorphism group of an order pq which is cyclic abelian group.
(2) If p divides $q - 1$, then the group of an order pq is not abelian group.
(3) Every group of an order 15, 35 or 185 is a cyclic group. This follows from the property (1)
(4) Every group of an order p^2 is an abelian one. So in this case there exist only two different up to isomorphism groups of an order p^2. These are cyclic group of an order p^2 and a group isomorphic to Cartesian product of two cyclic groups of an order p.
(5) A group of an order pq is solvable where p, q are different prime numbers.

Now we can generalize definitions introduced earlier.

Trough $\boldsymbol{A}_i \in N, i = 1, 2, \ldots$, we shall denote the elementary event which consists in the implementation of script which is an algebraic formalization $\boldsymbol{A}_i$ during the system S functioning.

The table

Algebraic formalization	$\boldsymbol{A}_1$	$\boldsymbol{A}_2$	…	$\boldsymbol{A}_k$	…
Probability	p_1	p_2	…	p_k	…

such that $\sum_{i=1}^{\infty} p_i = 1$, where $\boldsymbol{N} = \langle N, \vee \wedge \rangle$, where $N = \{\boldsymbol{A}_i = \langle A_i, \Omega_F^i, \Omega_P^i \rangle | i. = 1, 2, \ldots\}$, $\boldsymbol{A}_i \in N, i = 1, 2, \ldots$, is a complete distributive lattice of algebraic systems, is called the distribution row of the realizations of a system's scenarios of functioning if the lattice $\boldsymbol{N}$ is infinite (countable). Let us note that the top row of the table is ordered by the relation ρ.

If the lattice $\boldsymbol{N} = \langle N, \vee\wedge \rangle$ is finite, $N = \{\boldsymbol{A}_i = \langle A_i, \Omega^i_F, \Omega^i_P \rangle | i = 1, 2, \ldots k\}$ then the distribution row of the realization of a system's scenarios of functioning is called the table

Algebraic formalization	$\boldsymbol{A}_1$	$\boldsymbol{A}_2$	…	$\boldsymbol{A}_k$
Probability	p_1	p_2	…	p_k

such that $\sum_{i=1}^{k} p_i = 1$ and the top row of the table is ordered by the relation ρ.

Let us now consider the relationship of the notion of a final sustainability of a system with the classic notion of a sustainability of a system.

10.5 The Connection Between Notions of Final Sustainability, Stationary Points and Classical Sustainability

Let us reminder the classical notion of Lyapunov sustainability. We shall use definitions from [7].

Let us consider a discrete dynamical system defined by the map f:

$$u \mapsto f(u) = f(u, r), \quad u \in R, r \in R^m, \ f : R \to R \tag{10.2}$$

Definition 10.10 (*Lyapunov Sustainability*) [7]

A fixed point u^* of the map (10.2) is called a sustainable one if for every $\varepsilon > 0$ there exists such $\delta > 0$ that for any initial data u_0 from the δ—neighborhood of the point u^* the whole system's trajectory u_t, t = 0,1,2,3,… is contained in the ε—neighborhood of the point u^*.

Let N_t, $t \in T$, be an indicator describing the system S.

Suppose that there is a ratio

$$N_{t+1} = f(N_t), \quad N_t \in R, \ f : R \to R \tag{10.3}$$

or equivalently,

$$N \to f(N), \quad N \in R \tag{10.4}$$

where f is a given map, depending on a number of parameters.

Equations (10.3) and (10.4) defined the scalar discrete dynamical systems.

The following definitions are contained in [7].

Definition 10.11 The phase space or a state space of a system S is the set of all its states $\{N_t | t \in R\}$, (10.3).

Definition 10.12 A trajectory or an orbit of a system (10.2), generated by the map f is a sequences of points $\{N_t | t = 1, 2, \ldots\}$.

Definition 10.13 The points N^* from the set of states $\{N_t | t \in R\}$ of a system (10.3) such that $f(N^*) = N^*$ are called fixed points of the system (10.3).

The relations (10.3) and (10.4) specify one-dimensional discrete dynamical systems, so the state space of the system (10.3) is a subset of the real line or the additive group $\langle R, +, -, 0, \leq \rangle$. Fixed points which are defined by relations $f(N^*) = N^*$, are the simplest type of trajectories and they correspond to the stationary states of the system (10.3).

Herewith it is necessary to know the answer to the following question: if the system is in a sufficiently small neighborhood of the fixed point will it remain there over the time or will it be out of this neighborhood?

That is one must know if the fixed point is stable or not.

We have shown that every $a \in G_S$ defines the homomorphism $\breve{a} : T \to G_S \times T$, such that for any $x \in T$ one has $x \xrightarrow{\breve{a}} a(x)$. For a complete preimage of the set $M \subseteq G_S \times T$ under the map $\breve{a}$ the inclusion $(\breve{a})^{-1}(M) \subseteq T$ takes place. Every $x \in T$ defines the homomorphism $\widehat{x} : G_S \to G_S \times T$, such that for any $a \in G_S$ one has $a \xrightarrow{\widehat{x}} a(x)$. For a complete preimage of the set $M \subseteq G_S \times T$ under the map $\widehat{x}$ the inclusion $(\widehat{x})^{-1}(M) \subseteq G_S$ takes place.

Let f be a map from T into T setting the trajectory of the development of the system S. The indicator which characterizes the system S is the Cayley table of a group G_S in our case. This is a quality score or a complex index which contains basic quantitative factors determining the system S.

As it was mentioned earlier,

$$\begin{gathered} f : T \to T \\ \left((\breve{a})^{-1}(M)\right) \mapsto f\left((\breve{a})^{-1}(M)\right) \end{gathered} \tag{10.5}$$

Now suppose that for any time moment $x \in T$ we have $i^{-1} \propto^{-1} (A_n) \subseteq M$, where i is an imbedding $G_S \to Symm\, G_S \cong S_n$, and S_n is a symmetric group of permutations of a degree n, $\propto$ is an isomorphism of a group $Symm\, G_S$ onto S_n, $\propto$: $Symm\, G \to S_n$. Then by the Theorem 10.4 the system S will be a final sustainable one in the sense of the Definition 10.1 because its trajectory of development does not change, so

$$f\left((\breve{a})^{-1}(M)\right) = \left((\breve{a})^{-1}(M)\right)$$

and $(\breve{a})^{-1}(M)$ are fixed points of the system S. Thus, the system S will be fixed in the sense of [7]. If there exists a time moment $x \in T$ such that for every subset $M \subseteq G_S \times T$ the condition

$$i^{-1} \propto^{-1} (A_n) \not\subseteq M \tag{10.6}$$

takes place then such points will not be fixed in the sense of the equation $f(N^*) = N^*$.

In fact if the Eq. (10.5)

$$f : T \to T$$
$$\left((\breve{a})^{-1}(M)\right) \mapsto f\left((\breve{a})^{-1}(M)\right),$$

takes place then $\left((\breve{a})^{-1}(M)\right) \subseteq T$, but it is not so because of (10.6).

As we have mentioned before the indicator $K(S)$ which is characterizing the system S, is the Cayley table of a group G_S. We should note that this is a complex indicator which includes essentially all the quantitative indicators which characterized a closed associative system S with reverse effects or a feedback.

In order to characterize the dynamics of the quantitative indicators included into the complex indicator one can use the following topologies which are described in [8]:

(1) Let us use an Example 1.3.9 from [8]. Let (Y, τ) be a topological space and X be a non empty set. Let $f : X \to Y$ be a function from X into Y and $\tau_1 = \{f^{-1}(S) | S \in \tau\}$. Then τ_1 is a topology on X. Besides it if X is a non empty subset of a topological space (Y, τ), then the set $\tau_X = \{o \cap X | o \in \tau\}$ defines a topology on Y, which is induced by the topology τ.

Let $T = (R, \tau)$ be an Euclidean topological space on the real line, (respectively one can consider the topological space $(Q, \{o \cap Q | o \in \tau\})$ where Q is a set of all rational numbers). Let us consider the natural embedding $w : ((\breve{a})^{-1}(M)) \hookrightarrow T((\breve{a})^{-1}(M)) \subseteq T = R$ where $i^{-1} \propto^{-1} (A_n) \subseteq M$, or the topology on $\left((\breve{a})^{-1}(M)\right)$, which is induced by τ. Then $\tau_1' = \{w^{-1}(S) | S \in \tau\}$ is a topological space with an Euclidean topology.

(2) Let us consider the pointwise convergence topology on the set of all transformations Ω. The topology of pointwise convergence is the topology induced by the product topology. If $F = \{\varphi : X \to Y\}$ is a set of maps of the set X into a topological space Y then F is contained in the direct product $Y^X = \prod\{Y | x \in X\}$. The topology of point wise convergence is the topology induced by the product topology, that is the orientation $\{f_n(x), n \in D\}$ converges to $g(x)$ in each $x \in X$. A prebase B of the topology is formed by all the subsets of the form $\{f | f(x) \in U\}$, where x is any point from X and U is any open set in X.

Let us recall the definitions of the product topology and Euclidean topology. The product topology on the direct product $X_1 \times X_2 \times \cdots \times X_n$ of topological spaces $(X_1, \tau_1), (X_2, \tau_2), \ldots, (X_n, \tau_n)$ is the strongest topology on $X_1 \times X_2 \times \cdots \times X_n$ such that every projection $X_1 \times X_2 \times \cdots \times X_n \to X_i$ is continuous, [8].

Corollary 10.14 *If a system is final sustainable, then it is Lyapunov sustainable in the sense of* Definitions 10.10 and 10.13.

Proof In fact the indicator which characterizes a system *S*, is Cayley table of a group of factors G_S. The initial conditions is Cayley table of a group G_S which is unique for a final sustainable system.

Now we give the necessary definitions from [8].

Definition 10.15 [8] Let *f* be a map of a set *X* into itself. A point $x \in X$ is called a fixed one of the map *f*, if $f(x) = x$.

Definition 10.16 [8] Let (X, d) be a metric space, and *f* is a map from *X* into *X*. The map *f* is called a contraction map if there exists $r \in (0, 1)$ such that $d(f(x_1), f(x_2)) \le r d(x_1, x_2)$ for all $x_1, x_2 \in X$.

Proposition 10.17 [8] *Let f be a contraction map of a metric space* (X, d) *into itself. Then f is a continuous one.*

10.6 Practice Example. Algebraic Formalization as a Tool of Assertion the Sustainability of Ranking Systems of an Evaluation of Activities of Universities

As an example consider a way of formalizing a synthesis of a system by its decomposition with the usage the technique of the theory of extensions of Abelian groups and on this basis we examine the sustainability[1] of the ranking systems of evaluation the effectiveness of universities.

On the solutions of the problem of measurement of the effectiveness of the universities' activities many world ranking systems are directed, such as THE World University Rankings; QS World University Rankings; Academic Ranking of World Universities; Russian Global Universities Ranking GUR.

As it is well known, the first three of these ranking systems during the process of analysis and synthesis give the following decomposition into subsystems:

The World University Rankings contains 13 parameters with weights which are expressed in percentages from the total score on the following categories:

[1]Let's explain the notion of a sustainability of a system once more. The system is a sustainable one if at withdrawing it by the external effects from the state of equilibrium (rest) it returns to it after the cessation of external influences. From the point of view of an algebraic formalization it means that there are restrictions on the number of final states of the system [3–6].

I category "Teaching and learning environment"—30% upon parameters:

1. The survey of the scientific university staff's environments—15%;
2. The ratio of the number of professors and teaching staff to the number of students in high school—4.5%;
3. The ratio of the number of masters who completed a degree Ph.D., to the number of bachelors enrolled on masters—2.25%;
4. The ratio of the number of professors and teaching staff with Ph.D. degree to the total number of professor and teaching staff—6%;
5. The ratio of an income of an educational institution from the scientific research to the number of scientific staff—2.25%;

II category "Research—volume, income, reputation"—30% upon parameters:

6. Research reputation among similar universities—18%;
7. Income from research activities, correlated with the number of employees and normalized for purchasing power parity—6%;
8. Assessment of the environment for scientific research which is calculated as the ratio of the numbers of papers indexed in Thomson Reuters as scientific to the number of professors and teaching staff and then this ratio normalized;
9. Assessment of the environment for scientific development—6%.

III category "Citation, influence, authority"—30% upon parameters:

10. The category consists from one indicator an index of citing. The ranking does not include universities that published less than 200 papers per year.
11. Data are normalized to reflect the difference in the various scientific fields.

IV category "Income from production activities: innovations"—2.5% upon parameter:

12. Evaluation of knowledge transfer—innovations, consulting, invention.

V category "International image"—7.5% upon parameters:

13. The share of foreign undergraduate and graduate students in relation to local students and postgraduates—2.5%;
14. The share of foreign citizens among the professors and teaching staff in relation to the total number of professors and teaching staff—2.5%;
15. The share of scientific publications, which have at least one foreign co-author relative to the total number of publications of the University—2.5%.

There are five blocks in this ranking, so in accordance with the consequence 2.3 this ranking is a rather sustainable one.

QS World University Ranking uses the following indicators:

1. Expert survey in the scientific community—40%;
2. Employers survey—10%;
3. The ratio of a number of students to the number of professors and teacher staff—20%;
4. The number of citations per staff member of the teaching staff—20%;
5. The share of foreign students—5%;
6. The share of foreign employees among professors and teaching staff—5%. This indicator was added later.

The introduction of the sixth indicator makes a system less sustainable, Corollary 10.3

ARWU Academic Ranking of World Universities uses the following indicators combined into five groups:

1. Alumni (Holders of the Nobel Prize or Fields Medal, which graduated from the University) 10%,
2. Award (the total number of employees of the University, received the Nobel Prize or Fields Medal at the time of work at the University) 20%,
3. HiCi (the number of scientists which are the most highly cited in 21 subject areas) 20%,
4. N&S (the number of scientific articles published in the journals Nature and Science from 2008 to 201…), 20%
5. PUB (the total number of articles indexed in the databases Science Citation Index-Expanded и Social Science Citation, 20%,
6. PCP (five points of the preceding indicators, divided into an equivalent number of academic staff working full time) 10%. This indicator is a derivative one from the previous indicators.

There are five basic blocks, 1–5, in this ranking, so in accordance with the consequence 2.3 this ranking is a rather sustainable one.

Shanghai Jiaotong University make up global research profiles of universities **GRUP (Global Research University Profile)**, using the following indicators: 14 indicators for students, 9 for professors and teaching staff, 13 to describe the infrastructure and economy of the university and 5 for research, a total number of indicators is 41 which is a simple number. So in accordance with Corrollary 10.3 this ranking is rather sustainable.

Now let's remind the Theorem 4.8 from the Chap. 4 and the theorem from [9] about description the group of extensions:

Theorem 4.8, Chap. 4 *Let the operation of composition offactors which represent the closed associative system with a feedback be a commutative one. Then the synthesis of the systems* S_1, S_2 *is described by the group offactors* $Ext(B_2, B_1)$, [2]

[2]The group of extensions of an abelian group B_1 by the abelian group B_2, [9].

the synthesis of the systems S_1, S_2, S_3 *is described by the group of factors* $Ext(B_3, Ext(B_2, B_1))$, *the synthesis of the systems* S_1, S_2, S_3, S_4 *is described by the group of factors* $Ext(B_4, Ext(B_3, Ext(B_2, B_1)))$, $Ext(B_4, Ext(B_3, Ext(B_2, B_1)))$, *the synthesis of the systems* B_1, B_2, B_3, B_4, B_5 *is described by the group of factors* $Ext(B_5, Ext(B_4, Ext(B_3, Ext(B_2, B_1))))$.

As the numbers of factors representing a close associative system S with a feedback with commutative operation of composition of factors, is finite, then the following theorem [9], allows to clear up to some extent in this case, the synthesis process of the system S.

Theorem [9] *If A and C are finite groups then the group of extensions* $Ext(C, A)$ *is isomorphic to the group of all homomorphisms* $Hom(C, A)$, *that is* $Ext(C, A) \cong Hom(C, A)$.

For example, if $\boldsymbol{G_S} \cong \boldsymbol{Z_m}$ is a cyclic group of order m, then $Hom(\boldsymbol{Z_m}, \boldsymbol{Z_m}) \cong \boldsymbol{Z_m} \cdot Hom(\boldsymbol{Z_m}, \boldsymbol{Z_m}) \cong \boldsymbol{Z_m}$.

This construction, that is ranking system, can be used for building ranking systems monitoring smart universities. Herewith, the blocks of rating systems themselves will change, because in this case one will have to evaluate and compare:

- systems for monitoring the results of the educational process,
- expert communities,
- active educational technologies,
- modules of educational resources,
- quality of IT—technologies,
- system of formation of individual educational trajectories,
- technology to determine the personality characteristics of a student,
- the effectiveness of financial support for the activities of a smart university, and others.

This will help to create a monitoring of the education system that tracks the quality of education better than existing ranking systems of an evaluation of activities of universities.

Using both of these theorems and the tensor estimate of system's functioning considered in Chap. 6, we can construct new ranking system to monitor and to manage Smart Education System.

It is also important that, it will help to make Smart Education System more sustainable.

10.7 Conclusions

The obtained results allow to extract the following conclusions.

For smart university to take leading positions in the world ranking systems we should to ensure and we are providing now:

1. The development and the construction of optimization models for each group of indicators and, based on overall optimization model to plan reliably the development of the smart education system with the priority of sustainable development and continuity of smart scientific schools.
2. The construction of the model and the plan of the optimal realization of smart education on the base of ranking system for smart universities.
3. The consideration of the establishment of a new smart ranking system which has the assessment of sustainability greater than that of well—known foreign systems. The formation such a system will provide a competitive advantage in the evaluation of the quality of national smart educations.

With regard to the Smart University System the obtained results allow to formulate the following recommendations to ensure the effectiveness of this system. It is desirable the development and construction a model of optimization for each group of indicators of Smart University system efficiency and on this basis to build a common optimization model for reliably planning the development of Smart University education system with a priority to ensure sustainable development and continuity of smart scientific schools.

References

1. Demidovich, B.P.: Lectures on Mathematical Theory of Sustainability. Nauka, Moscow (1967)
2. Nogin, V.D.: Theory of Stability of Motion. Faculty of Applied Mathematics and Control Processes, St. Petersburg State University, St. Petersburg (2008). (in Russian)
3. Mesarovich, M., Takahara, Y.: General System Theory: Mathematical Foundations. Mathematics in Science and Engineering, vol. 113. Academic Press, New York, San Francisco, London (1975)
4. Chernova, N.I.: Probability Theory. Novosibirsk State University, Novosibirsk (2007). (in Russian)
5. Vavilov, N.: Specific Theory of Groups. LNCS Homepage http://pps.kaznu.kz/2/Main/FileShow/684223/89/124/3348/. (in Russian)
6. Kurosh, A.G.: Theory of Groups. Nauka, Moscow (1967). (in Russian)
7. Bratus, A.S., Novozhilov, A.S., Rodina, E.V.: Discrete Dynamical Systems and Models in Ecology. Moscow State University of Railway Engineering, Moscow (2005). (in Russian)
8. Morris, S.A.: Topology Without Tears, Version of October 4 (2014). LNCS Homepage http://www.topologywithouttears.net/topbook.pdf
9. Fuchs, L.: Infinite Abelian Groups, vol. 1, Mir. Moscow (1974). (in Russian)

Printed in the United States
By Bookmasters